Health Risk and Exposure Assessment for Ozone

Second External Review Draft

Chapter 6 Appendices

DISCLAIMER

This draft document has been prepared by staff from the Risk and Benefits Group, Health and Environmental Impacts Division, Office of Air Quality Planning and Standards, U.S. Environmental Protection Agency. Any findings and conclusions are those of the authors and do not necessarily reflect the views of the Agency. This draft document is being circulated to facilitate discussion with the Clean Air Scientific Advisory Committee to inform the EPA's consideration of the ozone National Ambient Air Quality Standards.

This information is distributed for the purposes of pre-dissemination peer review under applicable information quality guidelines. It has not been formally disseminated by EPA. It does not represent and should not be construed to represent any Agency determination or policy.

Questions related to this preliminary draft document should be addressed to Dr. Bryan Hubbell, U.S. Environmental Protection Agency, Office of Air Quality Planning and Standards, C539-07, Research Triangle Park, North Carolina 27711 (email: hubbell.bryan@epa.gov).

EPA-452/P-14-004d
February 2014

Health Risk and Exposure Assessment for Ozone
Second External Review Draft
Chapter 6 Appendices

U.S. Environmental Protection Agency
Office of Air and Radiation
Office of Air Quality Planning and Standards
Health and Environmental Impacts Division
Risk and Benefits Group
Research Triangle Park, North Carolina 27711

1

2

3

4 # Appendix 6A

5

6 # Probabilistic Population Exposure-Response Relationships

7

8   This appendix shows the probabilistic exposure-response relationships for lung function

9   decrements associated with 8-hour O_3 exposures occurring at moderate exertion. The 2.5[th]

10   percentile, median (50th percentile), and 97.5[th] percentile exposure-response functions for

11   changes in $FEV_1 \geq 10\%$, $\geq 15\%$ and $\geq 20\%$ are shown in Figures 6A-1 through 6A-3, along with

12   the response data to which they were fit. The values of the function are provided in Table 6A-1.

13

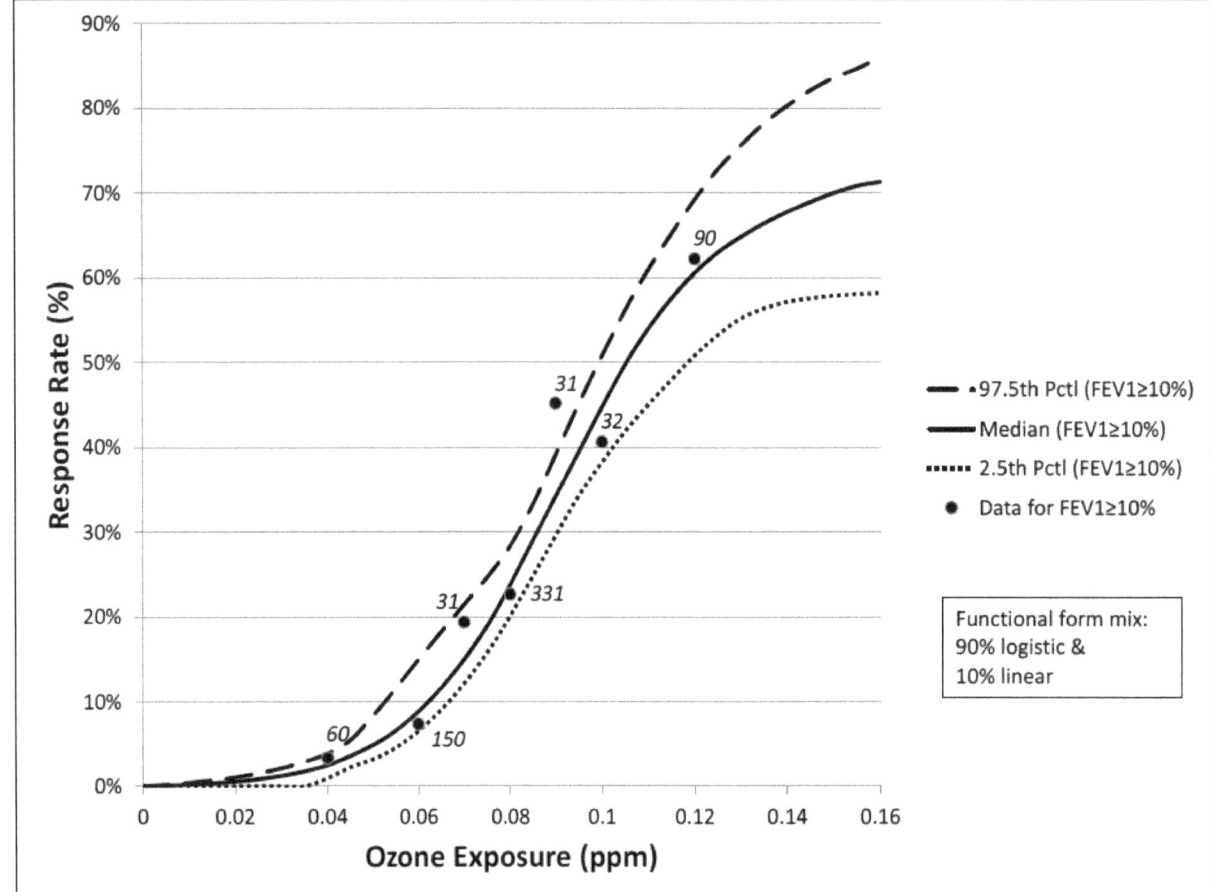

15   **Figure 6A-1. Probabilistic Exposure-Response Relationships for FEV1 Decrement ≥ 10%**
16   **for 8-Hour Exposures At Moderate Exertion, Ages 18-35. Values associated with data**
17   **points are the number of subject-exposures at each exposure concentration.**
18
19

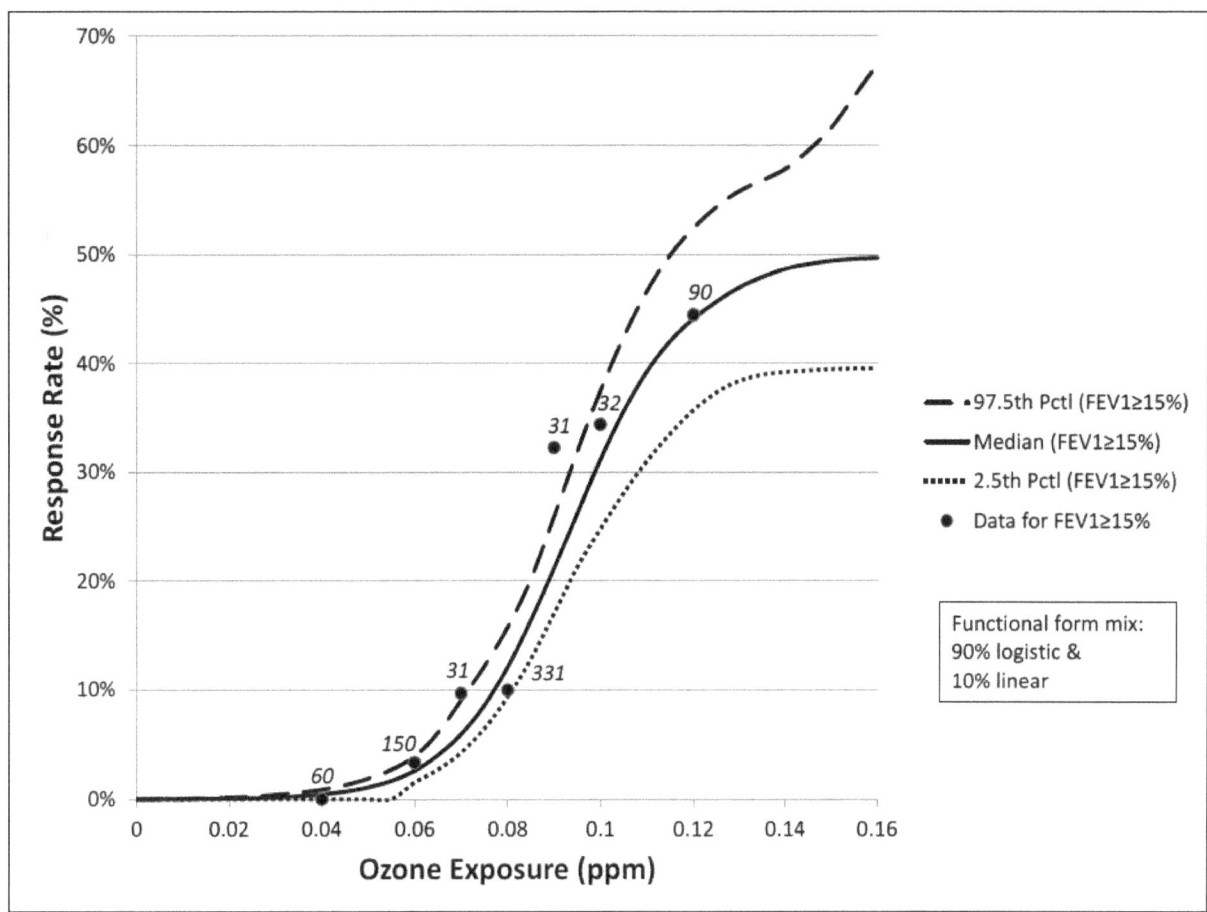

20

21 Figure 6A-2 Probabilistic Exposure-Response Relationships for FEV1 Decrement ≥ 15%
22 for 8-Hour Exposures At Moderate Exertion, Ages 18-35. Values associated with data
23 points are the number of subject-exposures at each exposure concentration.

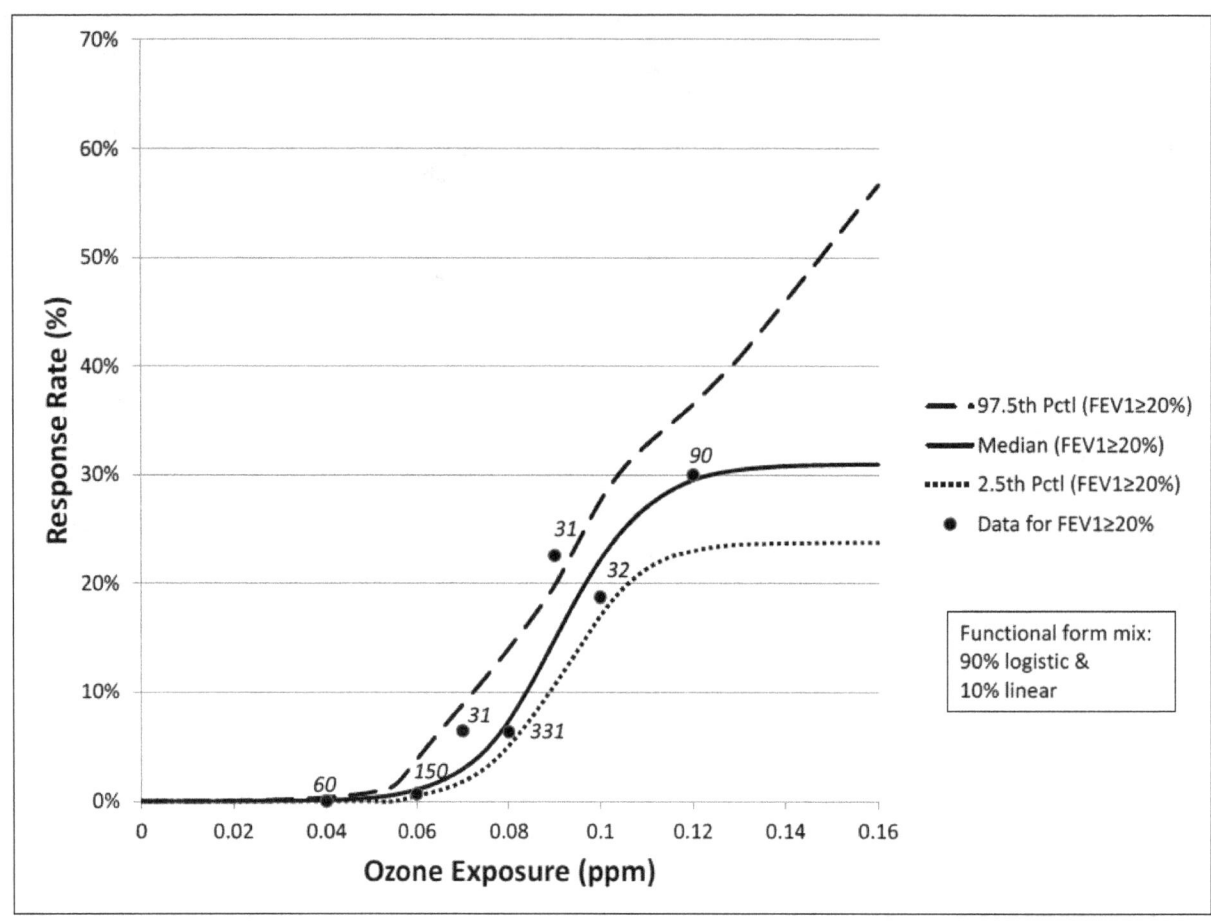

24

25 **Figure 6A-3. Probabilistic Exposure-Response Relationships for FEV1 Decrement ≥ 20%**
26 **for 8-Hour Exposures At Moderate Exertion, Ages 18-35. Values associated with data**
27 **points are the number of subject-exposures at each exposure concentration.**
28
29

30

31
32
33

Table 6A-1. Probabilistic Exposure-Response Relationships for FEV$_1$ Decrement $\geq$ 10%, $\geq$ 15%, and $\geq$ 20% for 6.6-Hour Exposures at Moderate Exertion with Functional Mix of 90% Logistic and 10% Linear Form, Ages 18-35

O$_3$ (ppm)	$\geq$ 10%			$\geq$ 15%			$\geq$ 20%		
	2.5%	median	97.5%	2.5%	median	97.5%	2.5%	median	97.5%
0	0	0	0	0	0	0	0	0	0
0.005	0	0.0008	0.0018	0	0.0001	0.0002	0	0	0.0001
0.01	0	0.0019	0.0041	0	0.0002	0.0006	0	0	0.0002
0.015	0	0.0035	0.0070	0	0.0004	0.0010	0	0.0001	0.0003
0.02	0	0.0056	0.0106	0	0.0007	0.0017	0	0.0001	0.0006
0.025	0	0.0084	0.0153	0	0.0011	0.0027	0	0.0002	0.0009
0.03	0	0.0123	0.0213	0	0.0018	0.0041	0	0.0003	0.0015
0.035	0	0.0176	0.0289	0	0.0029	0.0060	0	0.0006	0.0023
0.04	0.0095	0.0249	0.0389	0	0.0045	0.0088	0	0.0011	0.0036
0.045	0.0225	0.0362	0.0550	0	0.0070	0.0129	0	0.0019	0.0055
0.05	0.0325	0.0495	0.0848	0	0.0109	0.0188	0	0.0033	0.0084
0.055	0.0464	0.0665	0.1168	0	0.0167	0.0270	0	0.0060	0.0150
0.06	0.0653	0.0883	0.1492	0.0152	0.0260	0.0393	0.0052	0.0108	0.0383
0.065	0.0901	0.1160	0.1818	0.0271	0.0404	0.0609	0.0099	0.0180	0.0635
0.07	0.1213	0.1497	0.2146	0.0427	0.0595	0.0906	0.0180	0.0296	0.0887
0.075	0.1586	0.1905	0.2475	0.0647	0.0860	0.1223	0.0312	0.0476	0.1147
0.08	0.2010	0.2378	0.2835	0.0934	0.1212	0.1577	0.0508	0.0738	0.1404
0.085	0.2473	0.2894	0.3319	0.1287	0.1642	0.2018	0.0765	0.1083	0.1675
0.09	0.2951	0.3415	0.3907	0.1700	0.2115	0.2590	0.1069	0.1482	0.1981
0.095	0.3428	0.3948	0.4511	0.2121	0.2614	0.3181	0.1388	0.1879	0.2376
0.1	0.3825	0.4474	0.5079	0.2479	0.3116	0.3747	0.1703	0.2219	0.2762
0.105	0.4178	0.4961	0.5604	0.2810	0.3560	0.4242	0.1954	0.2493	0.3055
0.11	0.4496	0.5393	0.6089	0.3099	0.3922	0.4656	0.2134	0.2704	0.3277
0.115	0.4796	0.5756	0.6520	0.3349	0.4199	0.4990	0.2245	0.2853	0.3462
0.12	0.5080	0.6055	0.6921	0.3569	0.4408	0.5242	0.2299	0.2952	0.3650
0.125	0.5319	0.6292	0.7273	0.3737	0.4567	0.5439	0.2340	0.3012	0.3859
0.13	0.5517	0.6477	0.7557	0.3838	0.4695	0.5581	0.2362	0.3047	0.4082
0.135	0.5637	0.6639	0.7819	0.3895	0.4789	0.5677	0.2368	0.3068	0.4346
0.14	0.5715	0.6774	0.8026	0.3920	0.4867	0.5784	0.2373	0.3082	0.4603
0.145	0.5757	0.6893	0.8209	0.3934	0.4912	0.5957	0.2375	0.3089	0.4868
0.15	0.5786	0.6999	0.8356	0.3945	0.4941	0.6166	0.2378	0.3093	0.5138
0.155	0.5805	0.7084	0.8467	0.3951	0.4959	0.6449	0.2379	0.3096	0.5406
0.16	0.5819	0.7133	0.8603	0.3955	0.4968	0.6749	0.2379	0.3097	0.5668

34
35

1
2
3
Appendix 6B
4
5
Lung Function Risk Estimates Based on the MSS Model
6
7
8

13
14

Study area=Atlanta

year	scenario	FEV10 ages 5to18	FEV15 ages 5to18	FEV20 ages 5to18	FEV10 ages 19to35	FEV15 ages 19to35	FEV20 ages 19to35	FEV10 ages 36to55	FEV15 ages 36to55	FEV20 ages 36to55
2006	60 (06-08)	10.05%	1.66%	0.47%	2.66%	0.32%	0.07%	0.61%	0.03%	0.00%
2006	65 (06-08)	13.22%	2.75%	0.87%	3.62%	0.55%	0.13%	0.87%	0.06%	0.00%
2006	70 (06-08)	15.85%	3.80%	1.36%	4.52%	0.84%	0.19%	1.16%	0.08%	0.01%
2006	75 (06-08)	19.21%	5.29%	2.05%	5.77%	1.24%	0.34%	1.56%	0.15%	0.02%
2006	base	30.77%	11.72%	5.76%	10.76%	2.92%	1.22%	3.57%	0.50%	0.13%
2007	60 (06-08)	9.98%	1.77%	0.51%	2.59%	0.36%	0.07%	0.54%	0.01%	0.00%
2007	65 (06-08)	13.13%	2.72%	0.90%	3.60%	0.57%	0.14%	0.81%	0.05%	0.00%
2007	70 (06-08)	15.77%	3.80%	1.34%	4.47%	0.76%	0.22%	1.12%	0.08%	0.01%
2007	75 (06-08)	18.99%	5.18%	2.09%	5.73%	1.13%	0.34%	1.51%	0.12%	0.01%
2007	base	30.20%	11.32%	5.48%	10.65%	2.88%	1.13%	3.32%	0.47%	0.09%
2008	60 (06-08)	7.03%	0.90%	0.20%	1.85%	0.23%	0.05%	0.41%	0.02%	0.00%
2008	60 (08-10)	9.77%	1.70%	0.45%	2.59%	0.38%	0.09%	0.60%	0.04%	0.00%
2008	65 (06-08)	9.09%	1.51%	0.36%	2.44%	0.34%	0.08%	0.56%	0.04%	0.00%
2008	65 (08-10)	12.17%	2.48%	0.83%	3.40%	0.54%	0.14%	0.76%	0.07%	0.01%
2008	70 (06-08)	11.00%	2.09%	0.61%	3.01%	0.45%	0.12%	0.68%	0.06%	0.01%
2008	70 (08-10)	15.41%	3.76%	1.40%	4.56%	0.79%	0.24%	1.07%	0.10%	0.02%
2008	75 (06-08)	13.36%	2.97%	1.03%	3.81%	0.64%	0.19%	0.87%	0.08%	0.01%
2008	75 (08-10)	19.22%	5.55%	2.21%	5.75%	1.17%	0.36%	1.53%	0.17%	0.03%
2008	base	22.62%	7.22%	3.15%	7.00%	1.55%	0.52%	1.97%	0.24%	0.04%
2009	60 (08-10)	7.36%	1.10%	0.29%	1.92%	0.26%	0.06%	0.39%	0.02%	0.00%
2009	65 (08-10)	9.29%	1.66%	0.49%	2.50%	0.37%	0.09%	0.52%	0.04%	0.00%
2009	70 (08-10)	11.88%	2.48%	0.84%	3.24%	0.52%	0.14%	0.71%	0.05%	0.01%
2009	75 (08-10)	14.94%	3.69%	1.34%	4.12%	0.73%	0.23%	1.00%	0.09%	0.02%
2009	base	17.36%	4.89%	1.92%	4.82%	0.97%	0.30%	1.25%	0.13%	0.02%
2010	60 (08-10)	9.84%	1.56%	0.40%	2.47%	0.32%	0.06%	0.52%	0.03%	0.01%
2010	65 (08-10)	12.10%	2.23%	0.70%	3.11%	0.43%	0.10%	0.69%	0.04%	0.01%
2010	70 (08-10)	14.86%	3.25%	1.11%	3.98%	0.61%	0.18%	1.00%	0.07%	0.01%
2010	75 (08-10)	18.09%	4.55%	1.66%	4.98%	0.85%	0.26%	1.31%	0.11%	0.01%
2010	base	20.59%	5.73%	2.24%	5.84%	1.11%	0.36%	1.58%	0.14%	0.02%

1
2

year	scenario	FEV10 ages 5to18	FEV15 ages 5to18	FEV20 ages 5to18	FEV10 ages 19to35	FEV15 ages 19to35	FEV20 ages 19to35	FEV10 ages 36to55	FEV15 ages 36to55	FEV20 ages 36to55
2006	60 (06-08)	9.60%	1.64%	0.45%	2.41%	0.34%	0.09%	0.46%	0.03%	0.01%
2006	65 (06-08)	12.75%	2.82%	0.86%	3.29%	0.52%	0.15%	0.70%	0.07%	0.01%
2006	70 (06-08)	15.62%	3.93%	1.38%	4.23%	0.73%	0.23%	0.96%	0.10%	0.02%
2006	75 (06-08)	18.62%	5.16%	2.05%	5.27%	1.03%	0.34%	1.27%	0.13%	0.02%
2006	base	27.12%	9.74%	4.58%	8.72%	2.17%	0.77%	2.37%	0.32%	0.08%
2007	60 (06-08)	9.61%	1.64%	0.43%	2.18%	0.29%	0.08%	0.47%	0.03%	0.00%
2007	65 (06-08)	12.25%	2.53%	0.81%	3.01%	0.43%	0.13%	0.66%	0.04%	0.00%
2007	70 (06-08)	14.52%	3.42%	1.18%	3.77%	0.59%	0.18%	0.87%	0.06%	0.01%
2007	75 (06-08)	16.95%	4.45%	1.68%	4.54%	0.80%	0.23%	1.12%	0.08%	0.02%
2007	base	23.30%	7.76%	3.42%	6.74%	1.53%	0.51%	1.90%	0.23%	0.04%
2008	60 (06-08)	7.40%	1.09%	0.26%	1.77%	0.22%	0.05%	0.32%	0.02%	0.00%
2008	60 (08-10)	7.92%	1.24%	0.32%	1.93%	0.25%	0.06%	0.35%	0.02%	0.00%
2008	65 (06-08)	9.55%	1.76%	0.55%	2.33%	0.31%	0.08%	0.46%	0.03%	0.01%
2008	65 (08-10)	9.83%	1.87%	0.59%	2.43%	0.32%	0.09%	0.47%	0.03%	0.01%
2008	70 (06-08)	11.45%	2.44%	0.80%	2.90%	0.40%	0.13%	0.59%	0.05%	0.01%
2008	70 (08-10)	11.95%	2.62%	0.88%	3.05%	0.45%	0.14%	0.62%	0.06%	0.01%
2008	75 (06-08)	13.38%	3.29%	1.18%	3.50%	0.58%	0.17%	0.75%	0.07%	0.01%
2008	75 (08-10)	13.85%	3.50%	1.27%	3.65%	0.62%	0.19%	0.81%	0.07%	0.01%
2008	base	19.29%	6.03%	2.69%	5.38%	1.16%	0.37%	1.43%	0.14%	0.03%
2009	60 (08-10)	6.74%	0.84%	0.17%	1.63%	0.22%	0.06%	0.31%	0.01%	0.00%
2009	65 (08-10)	8.16%	1.24%	0.30%	1.97%	0.28%	0.07%	0.40%	0.02%	0.00%
2009	70 (08-10)	9.71%	1.78%	0.45%	2.37%	0.35%	0.10%	0.51%	0.03%	0.00%
2009	75 (08-10)	11.07%	2.23%	0.66%	2.79%	0.43%	0.12%	0.60%	0.05%	0.01%
2009	base	14.55%	3.81%	1.36%	3.75%	0.67%	0.21%	0.86%	0.07%	0.01%
2010	60 (08-10)	10.61%	1.95%	0.50%	2.66%	0.32%	0.10%	0.49%	0.04%	0.00%
2010	65 (08-10)	13.00%	2.78%	0.92%	3.40%	0.47%	0.15%	0.68%	0.06%	0.00%
2010	70 (08-10)	15.65%	3.88%	1.39%	4.25%	0.72%	0.21%	0.91%	0.09%	0.01%
2010	75 (08-10)	18.09%	4.99%	1.97%	5.17%	0.95%	0.31%	1.17%	0.11%	0.02%
2010	base	24.32%	8.27%	3.71%	7.50%	1.67%	0.62%	2.03%	0.25%	0.06%

3
4

Study area=Boston

year	scenario	FEV10 ages 5to18	FEV15 ages 5to18	FEV20 ages 5to18	FEV10 ages 19to35	FEV15 ages 19to35	FEV20 ages 19to35	FEV10 ages 36to55	FEV15 ages 36to55	FEV20 ages 36to55
2006	60 (06-08)	6.98%	1.06%	0.26%	1.73%	0.21%	0.05%	0.36%	0.02%	0.00%
2006	65 (06-08)	8.99%	1.75%	0.47%	2.33%	0.31%	0.09%	0.56%	0.03%	0.00%
2006	70 (06-08)	11.79%	2.83%	0.93%	3.06%	0.55%	0.15%	0.83%	0.06%	0.01%
2006	75 (06-08)	13.63%	3.71%	1.39%	3.59%	0.72%	0.21%	1.03%	0.09%	0.02%
2006	base	17.25%	5.51%	2.40%	4.76%	1.05%	0.35%	1.42%	0.17%	0.03%
2007	60 (06-08)	8.88%	1.70%	0.45%	2.02%	0.24%	0.05%	0.45%	0.04%	0.01%
2007	65 (06-08)	11.57%	2.77%	0.88%	2.78%	0.41%	0.09%	0.69%	0.06%	0.01%
2007	70 (06-08)	15.17%	4.42%	1.67%	3.87%	0.69%	0.20%	1.12%	0.10%	0.03%
2007	75 (06-08)	17.75%	5.69%	2.36%	4.65%	0.97%	0.28%	1.40%	0.15%	0.04%
2007	base	21.88%	7.82%	3.77%	5.94%	1.39%	0.46%	1.96%	0.24%	0.06%
2008	60 (06-08)	6.67%	0.95%	0.19%	1.73%	0.21%	0.04%	0.37%	0.02%	0.00%
2008	60 (08-10)	7.67%	1.29%	0.29%	1.94%	0.25%	0.06%	0.46%	0.03%	0.00%
2008	65 (06-08)	8.55%	1.59%	0.44%	2.20%	0.30%	0.07%	0.55%	0.03%	0.00%
2008	65 (08-10)	10.25%	2.20%	0.68%	2.64%	0.42%	0.10%	0.70%	0.04%	0.00%
2008	70 (06-08)	11.14%	2.53%	0.83%	2.84%	0.48%	0.12%	0.76%	0.05%	0.00%
2008	70 (08-10)	12.90%	3.20%	1.16%	3.35%	0.62%	0.17%	0.95%	0.07%	0.01%
2008	75 (06-08)	12.90%	3.20%	1.16%	3.35%	0.62%	0.17%	0.95%	0.07%	0.01%
2008	75 (08-10)	14.95%	4.26%	1.72%	3.95%	0.80%	0.25%	1.16%	0.10%	0.02%
2008	base	15.75%	4.74%	1.94%	4.13%	0.86%	0.28%	1.27%	0.12%	0.03%
2009	60 (08-10)	7.38%	1.36%	0.37%	1.80%	0.24%	0.06%	0.36%	0.02%	0.00%
2009	65 (08-10)	9.43%	2.06%	0.64%	2.28%	0.33%	0.10%	0.53%	0.03%	0.01%
2009	70 (08-10)	11.18%	2.83%	1.04%	2.71%	0.45%	0.15%	0.71%	0.05%	0.01%
2009	75 (08-10)	12.09%	3.26%	1.23%	2.87%	0.53%	0.17%	0.80%	0.05%	0.01%
2009	base	12.39%	3.41%	1.29%	2.89%	0.55%	0.16%	0.82%	0.06%	0.01%
2010	60 (08-10)	7.98%	1.24%	0.26%	1.94%	0.23%	0.05%	0.42%	0.03%	0.00%
2010	65 (08-10)	10.39%	2.05%	0.58%	2.52%	0.34%	0.07%	0.63%	0.04%	0.01%
2010	70 (08-10)	12.77%	3.04%	0.97%	3.18%	0.50%	0.12%	0.84%	0.06%	0.01%
2010	75 (08-10)	14.39%	3.68%	1.38%	3.54%	0.58%	0.14%	0.98%	0.07%	0.01%
2010	base	14.90%	3.97%	1.51%	3.66%	0.60%	0.15%	1.03%	0.08%	0.01%

Study area=Chicago

year	scenario	FEV10 ages 5to18	FEV15 ages 5to18	FEV20 ages 5to18	FEV10 ages 19to35	FEV15 ages 19to35	FEV20 ages 19to35	FEV10 ages 36to55	FEV15 ages 36to55	FEV20 ages 36to55
2006	60 (06-08)	8.23%	1.47%	0.42%	2.21%	0.31%	0.08%	0.45%	0.03%	0.00%
2006	65 (06-08)	10.45%	2.20%	0.72%	2.79%	0.47%	0.11%	0.62%	0.05%	0.01%
2006	70 (06-08)	12.54%	3.03%	1.13%	3.35%	0.59%	0.18%	0.79%	0.08%	0.02%
2006	75 (06-08)	14.39%	3.94%	1.57%	3.75%	0.75%	0.23%	0.96%	0.11%	0.03%
2006	base	14.55%	4.15%	1.65%	3.71%	0.75%	0.22%	0.96%	0.11%	0.03%
2007	60 (06-08)	10.75%	2.17%	0.64%	2.76%	0.40%	0.12%	0.61%	0.03%	0.00%
2007	65 (06-08)	13.86%	3.39%	1.19%	3.73%	0.65%	0.19%	0.94%	0.07%	0.01%
2007	70 (06-08)	16.95%	4.72%	1.81%	4.67%	0.90%	0.27%	1.24%	0.10%	0.02%
2007	75 (06-08)	19.68%	6.16%	2.61%	5.49%	1.14%	0.40%	1.59%	0.15%	0.03%
2007	base	20.14%	6.48%	2.81%	5.55%	1.18%	0.41%	1.64%	0.16%	0.03%
2008	60 (06-08)	7.07%	1.01%	0.24%	1.89%	0.22%	0.06%	0.35%	0.02%	0.00%
2008	60 (08-10)	8.25%	1.34%	0.33%	2.18%	0.25%	0.08%	0.42%	0.02%	0.00%
2008	65 (06-08)	8.69%	1.46%	0.36%	2.30%	0.28%	0.09%	0.47%	0.02%	0.00%
2008	65 (08-10)	9.85%	1.85%	0.51%	2.56%	0.33%	0.10%	0.56%	0.04%	0.00%
2008	70 (06-08)	10.14%	1.95%	0.54%	2.62%	0.35%	0.11%	0.58%	0.04%	0.00%
2008	70 (08-10)	11.14%	2.40%	0.78%	2.78%	0.39%	0.11%	0.66%	0.04%	0.01%
2008	75 (06-08)	11.17%	2.41%	0.80%	2.76%	0.38%	0.11%	0.65%	0.04%	0.01%
2008	75 (08-10)	11.05%	2.43%	0.81%	2.65%	0.37%	0.11%	0.65%	0.04%	0.00%
2008	base	11.05%	2.43%	0.81%	2.65%	0.37%	0.11%	0.65%	0.04%	0.00%
2009	60 (08-10)	8.16%	1.45%	0.42%	2.31%	0.34%	0.09%	0.46%	0.03%	0.00%
2009	65 (08-10)	9.86%	2.04%	0.63%	2.79%	0.42%	0.13%	0.59%	0.04%	0.01%
2009	70 (08-10)	11.31%	2.74%	0.89%	3.09%	0.52%	0.15%	0.70%	0.05%	0.01%
2009	75 (08-10)	11.30%	2.79%	0.92%	2.91%	0.51%	0.14%	0.67%	0.05%	0.01%
2009	base	11.30%	2.79%	0.92%	2.91%	0.51%	0.14%	0.67%	0.05%	0.01%
2010	60 (08-10)	10.83%	2.11%	0.60%	3.13%	0.51%	0.15%	0.66%	0.04%	0.01%
2010	65 (08-10)	13.19%	3.13%	1.02%	3.83%	0.71%	0.20%	0.88%	0.06%	0.01%
2010	70 (08-10)	15.34%	4.16%	1.46%	4.39%	0.88%	0.26%	1.09%	0.09%	0.02%
2010	75 (08-10)	15.60%	4.41%	1.64%	4.25%	0.88%	0.25%	1.07%	0.09%	0.02%
2010	base	15.60%	4.41%	1.64%	4.25%	0.88%	0.25%	1.07%	0.09%	0.02%

Study area=Cleveland

year	scenario	FEV10 ages 5to18	FEV15 ages 5to18	FEV20 ages 5to18	FEV10 ages 19to35	FEV15 ages 19to35	FEV20 ages 19to35	FEV10 ages 36to55	FEV15 ages 36to55	FEV20 ages 36to55
2006	60 (06-08)	4.40%	0.45%	0.06%	1.16%	0.10%	0.01%	0.25%	0.01%	0.00%
2006	65 (06-08)	7.68%	1.28%	0.30%	2.15%	0.29%	0.04%	0.50%	0.03%	0.01%
2006	70 (06-08)	10.55%	2.15%	0.66%	3.05%	0.49%	0.11%	0.73%	0.07%	0.01%
2006	75 (06-08)	13.45%	3.26%	1.17%	4.03%	0.78%	0.20%	0.99%	0.11%	0.02%
2006	base	17.33%	5.16%	2.12%	5.57%	1.23%	0.40%	1.47%	0.19%	0.05%
2007	60 (06-08)	5.42%	0.68%	0.17%	1.23%	0.12%	0.02%	0.27%	0.01%	0.00%
2007	65 (06-08)	9.95%	1.84%	0.58%	2.49%	0.31%	0.09%	0.55%	0.04%	0.01%
2007	70 (06-08)	13.59%	3.19%	1.10%	3.69%	0.53%	0.16%	0.92%	0.09%	0.01%
2007	75 (06-08)	17.02%	4.61%	1.84%	5.02%	0.91%	0.28%	1.32%	0.15%	0.02%
2007	base	21.65%	7.09%	3.19%	6.78%	1.50%	0.51%	2.07%	0.24%	0.05%
2008	60 (06-08)	4.79%	0.48%	0.07%	1.19%	0.12%	0.03%	0.24%	0.01%	0.00%
2008	60 (08-10)	4.79%	0.48%	0.07%	1.19%	0.12%	0.03%	0.24%	0.01%	0.00%
2008	65 (06-08)	8.54%	1.37%	0.38%	2.18%	0.28%	0.07%	0.52%	0.03%	0.00%
2008	65 (08-10)	7.50%	1.12%	0.25%	1.88%	0.22%	0.05%	0.42%	0.02%	0.00%
2008	70 (06-08)	11.58%	2.40%	0.79%	3.09%	0.50%	0.12%	0.77%	0.05%	0.01%
2008	70 (08-10)	10.88%	2.10%	0.69%	2.86%	0.44%	0.11%	0.70%	0.04%	0.01%
2008	75 (06-08)	14.50%	3.74%	1.34%	4.08%	0.74%	0.20%	1.08%	0.10%	0.01%
2008	75 (08-10)	14.10%	3.55%	1.27%	3.95%	0.69%	0.18%	1.03%	0.09%	0.01%
2008	base	18.32%	5.53%	2.25%	5.38%	1.13%	0.34%	1.54%	0.16%	0.03%
2009	60 (08-10)	4.31%	0.56%	0.09%	1.14%	0.10%	0.01%	0.20%	0.01%	0.00%
2009	65 (08-10)	6.15%	1.00%	0.24%	1.57%	0.20%	0.03%	0.31%	0.02%	0.00%
2009	70 (08-10)	8.45%	1.58%	0.49%	2.21%	0.33%	0.06%	0.50%	0.04%	0.00%
2009	75 (08-10)	10.58%	2.31%	0.78%	2.90%	0.49%	0.12%	0.69%	0.06%	0.00%
2009	base	12.83%	3.27%	1.21%	3.55%	0.66%	0.19%	0.88%	0.09%	0.02%
2010	60 (08-10)	5.11%	0.48%	0.09%	1.28%	0.11%	0.01%	0.24%	0.01%	0.00%
2010	65 (08-10)	7.79%	1.04%	0.23%	2.00%	0.23%	0.04%	0.44%	0.02%	0.00%
2010	70 (08-10)	11.10%	2.07%	0.57%	3.00%	0.43%	0.11%	0.69%	0.05%	0.01%
2010	75 (08-10)	14.31%	3.32%	1.07%	4.13%	0.68%	0.18%	1.01%	0.08%	0.01%
2010	base	18.49%	5.13%	2.08%	5.44%	1.15%	0.34%	1.47%	0.14%	0.03%

Study area=Dallas

year	scenario	FEV10 ages 5to18	FEV15 ages 5to18	FEV20 ages 5to18	FEV10 ages 19to35	FEV15 ages 19to35	FEV20 ages 19to35	FEV10 ages 36to55	FEV15 ages 36to55	FEV20 ages 36to55
2006	60 (06-08)	12.69%	2.60%	0.86%	3.43%	0.45%	0.10%	0.89%	0.06%	0.00%
2006	65 (06-08)	15.63%	3.76%	1.38%	4.44%	0.66%	0.18%	1.20%	0.10%	0.01%
2006	70 (06-08)	18.94%	5.11%	2.01%	5.73%	0.96%	0.29%	1.64%	0.15%	0.03%
2006	75 (06-08)	21.60%	6.39%	2.68%	6.93%	1.28%	0.41%	2.03%	0.22%	0.05%
2006	base	29.97%	10.90%	5.33%	10.54%	2.69%	0.92%	3.59%	0.48%	0.12%
2007	60 (06-08)	8.72%	1.51%	0.42%	2.16%	0.27%	0.06%	0.50%	0.03%	0.00%
2007	65 (06-08)	10.83%	2.20%	0.70%	2.79%	0.37%	0.09%	0.68%	0.05%	0.00%
2007	70 (06-08)	13.22%	3.10%	1.10%	3.53%	0.54%	0.14%	0.88%	0.08%	0.01%
2007	75 (06-08)	15.13%	3.91%	1.54%	4.17%	0.68%	0.19%	1.09%	0.11%	0.02%
2007	base	21.44%	7.21%	3.28%	6.43%	1.48%	0.50%	1.92%	0.25%	0.06%
2008	60 (06-08)	8.45%	1.31%	0.31%	2.16%	0.27%	0.04%	0.51%	0.02%	0.00%
2008	60 (08-10)	8.45%	1.31%	0.31%	2.16%	0.27%	0.04%	0.51%	0.02%	0.00%
2008	65 (06-08)	10.48%	1.92%	0.55%	2.69%	0.37%	0.08%	0.66%	0.03%	0.00%
2008	65 (08-10)	10.48%	1.92%	0.55%	2.69%	0.37%	0.08%	0.66%	0.03%	0.00%
2008	70 (06-08)	12.69%	2.69%	0.86%	3.37%	0.49%	0.13%	0.85%	0.06%	0.01%
2008	70 (08-10)	12.69%	2.69%	0.86%	3.37%	0.49%	0.13%	0.85%	0.06%	0.01%
2008	75 (06-08)	14.39%	3.46%	1.17%	3.97%	0.64%	0.19%	1.02%	0.07%	0.01%
2008	75 (08-10)	14.90%	3.64%	1.26%	4.14%	0.67%	0.20%	1.08%	0.08%	0.01%
2008	base	19.64%	6.01%	2.52%	5.85%	1.16%	0.39%	1.64%	0.17%	0.03%
2009	60 (08-10)	9.55%	1.77%	0.48%	2.40%	0.33%	0.08%	0.56%	0.03%	0.00%
2009	65 (08-10)	11.84%	2.53%	0.84%	3.13%	0.45%	0.13%	0.77%	0.05%	0.00%
2009	70 (08-10)	14.38%	3.48%	1.30%	3.98%	0.67%	0.19%	1.02%	0.09%	0.01%
2009	75 (08-10)	17.00%	4.64%	1.89%	5.01%	0.90%	0.28%	1.39%	0.12%	0.02%
2009	base	23.14%	7.85%	3.58%	7.32%	1.61%	0.60%	2.21%	0.27%	0.05%
2010	60 (08-10)	8.44%	1.31%	0.34%	2.13%	0.25%	0.05%	0.47%	0.02%	0.00%
2010	65 (08-10)	10.23%	1.88%	0.58%	2.71%	0.36%	0.07%	0.65%	0.05%	0.00%
2010	70 (08-10)	12.36%	2.61%	0.89%	3.39%	0.49%	0.11%	0.82%	0.07%	0.01%
2010	75 (08-10)	14.56%	3.52%	1.23%	4.10%	0.67%	0.17%	1.05%	0.11%	0.01%
2010	base	19.18%	5.82%	2.44%	5.50%	1.13%	0.38%	1.63%	0.18%	0.04%

Study area=Denver

year	scenario	FEV10 ages 5to18	FEV15 ages 5to18	FEV20 ages 5to18	FEV10 ages 19to35	FEV15 ages 19to35	FEV20 ages 19to35	FEV10 ages 36to55	FEV15 ages 36to55	FEV20 ages 36to55
2006	60 (06-08)	12.47%	2.31%	0.72%	3.77%	0.54%	0.12%	0.96%	0.07%	0.00%
2006	65 (06-08)	15.13%	3.32%	1.11%	4.80%	0.78%	0.17%	1.30%	0.10%	0.01%
2006	70 (06-08)	17.78%	4.49%	1.63%	5.74%	1.04%	0.29%	1.60%	0.16%	0.03%
2006	75 (06-08)	20.23%	5.64%	2.21%	6.68%	1.30%	0.40%	1.92%	0.22%	0.05%
2006	base	24.08%	7.75%	3.46%	8.12%	1.93%	0.66%	2.53%	0.38%	0.09%
2007	60 (06-08)	11.41%	1.96%	0.57%	3.40%	0.47%	0.10%	0.82%	0.05%	0.01%
2007	65 (06-08)	13.77%	2.83%	0.88%	4.22%	0.67%	0.14%	1.05%	0.09%	0.01%
2007	70 (06-08)	16.12%	3.68%	1.30%	4.95%	0.82%	0.21%	1.28%	0.12%	0.02%
2007	75 (06-08)	18.24%	4.57%	1.75%	5.67%	1.04%	0.30%	1.53%	0.16%	0.03%
2007	base	21.33%	6.31%	2.57%	6.68%	1.34%	0.42%	1.90%	0.25%	0.05%
2008	60 (06-08)	12.04%	2.14%	0.62%	3.79%	0.53%	0.12%	0.94%	0.05%	0.01%
2008	60 (08-10)	12.04%	2.14%	0.62%	3.79%	0.53%	0.12%	0.94%	0.05%	0.01%
2008	65 (06-08)	14.72%	3.05%	0.97%	4.77%	0.74%	0.18%	1.22%	0.09%	0.01%
2008	65 (08-10)	17.60%	4.04%	1.42%	5.82%	0.96%	0.30%	1.55%	0.14%	0.02%
2008	70 (06-08)	17.42%	3.97%	1.39%	5.76%	0.93%	0.29%	1.53%	0.13%	0.02%
2008	70 (08-10)	20.46%	5.31%	2.09%	7.01%	1.30%	0.42%	1.91%	0.20%	0.04%
2008	75 (06-08)	19.77%	4.96%	1.94%	6.71%	1.22%	0.39%	1.81%	0.19%	0.03%
2008	75 (08-10)	22.50%	6.46%	2.61%	7.77%	1.55%	0.51%	2.19%	0.25%	0.05%
2008	base	23.00%	6.78%	2.79%	7.94%	1.64%	0.52%	2.26%	0.26%	0.05%
2009	60 (08-10)	9.52%	1.52%	0.37%	2.87%	0.36%	0.09%	0.68%	0.03%	0.00%
2009	65 (08-10)	13.48%	2.75%	0.88%	4.23%	0.65%	0.16%	1.07%	0.07%	0.00%
2009	70 (08-10)	15.45%	3.58%	1.25%	4.89%	0.79%	0.20%	1.28%	0.10%	0.01%
2009	75 (08-10)	16.90%	4.35%	1.58%	5.23%	0.90%	0.24%	1.43%	0.13%	0.01%
2009	base	17.25%	4.55%	1.71%	5.36%	0.94%	0.26%	1.45%	0.13%	0.01%
2010	60 (08-10)	10.64%	1.73%	0.41%	3.18%	0.42%	0.10%	0.73%	0.05%	0.00%
2010	65 (08-10)	15.12%	3.19%	1.04%	4.71%	0.72%	0.21%	1.15%	0.11%	0.01%
2010	70 (08-10)	17.54%	4.11%	1.49%	5.58%	0.99%	0.28%	1.43%	0.13%	0.02%
2010	75 (08-10)	19.34%	5.01%	1.95%	6.14%	1.16%	0.34%	1.65%	0.17%	0.03%
2010	base	19.83%	5.31%	2.12%	6.25%	1.23%	0.37%	1.72%	0.17%	0.04%

Study area=Detroit

year	scenario	FEV10 ages 5to18	FEV15 ages 5to18	FEV20 ages 5to18	FEV10 ages 19to35	FEV15 ages 19to35	FEV20 ages 19to35	FEV10 ages 36to55	FEV15 ages 36to55	FEV20 ages 36to55
2006	60 (06-08)	7.49%	1.16%	0.28%	2.03%	0.28%	0.06%	0.47%	0.02%	0.00%
2006	65 (06-08)	9.57%	1.86%	0.51%	2.66%	0.42%	0.11%	0.63%	0.05%	0.01%
2006	70 (06-08)	11.37%	2.52%	0.81%	3.21%	0.54%	0.16%	0.79%	0.07%	0.01%
2006	75 (06-08)	13.64%	3.51%	1.28%	3.98%	0.77%	0.25%	1.00%	0.10%	0.02%
2006	base	17.53%	5.40%	2.33%	5.21%	1.19%	0.41%	1.38%	0.18%	0.04%
2007	60 (06-08)	9.07%	1.54%	0.42%	2.06%	0.29%	0.06%	0.47%	0.03%	0.01%
2007	65 (06-08)	12.04%	2.63%	0.79%	2.88%	0.44%	0.12%	0.71%	0.06%	0.01%
2007	70 (06-08)	14.54%	3.58%	1.24%	3.68%	0.63%	0.19%	0.95%	0.09%	0.02%
2007	75 (06-08)	17.79%	5.08%	2.07%	4.83%	0.94%	0.28%	1.31%	0.13%	0.03%
2007	base	23.67%	8.42%	3.88%	7.01%	1.65%	0.54%	2.15%	0.26%	0.06%
2008	60 (06-08)	7.05%	0.96%	0.20%	1.81%	0.22%	0.05%	0.41%	0.03%	0.00%
2008	60 (08-10)	8.44%	1.38%	0.35%	2.20%	0.31%	0.06%	0.50%	0.04%	0.00%
2008	65 (06-08)	8.89%	1.52%	0.40%	2.35%	0.34%	0.08%	0.53%	0.05%	0.00%
2008	65 (08-10)	11.07%	2.34%	0.70%	2.97%	0.48%	0.13%	0.68%	0.07%	0.01%
2008	70 (06-08)	10.46%	2.12%	0.59%	2.79%	0.43%	0.12%	0.63%	0.06%	0.01%
2008	70 (08-10)	13.70%	3.40%	1.21%	3.74%	0.67%	0.20%	0.95%	0.09%	0.02%
2008	75 (06-08)	12.68%	2.98%	0.99%	3.43%	0.60%	0.17%	0.83%	0.08%	0.01%
2008	75 (08-10)	15.80%	4.44%	1.78%	4.22%	0.86%	0.28%	1.14%	0.13%	0.02%
2008	base	15.80%	4.44%	1.78%	4.22%	0.86%	0.28%	1.14%	0.13%	0.02%
2009	60 (08-10)	7.31%	1.18%	0.28%	2.02%	0.27%	0.06%	0.43%	0.03%	0.00%
2009	65 (08-10)	9.49%	1.92%	0.57%	2.66%	0.43%	0.10%	0.61%	0.04%	0.01%
2009	70 (08-10)	11.50%	2.65%	0.94%	3.33%	0.58%	0.16%	0.81%	0.06%	0.01%
2009	75 (08-10)	13.11%	3.49%	1.39%	3.62%	0.70%	0.20%	0.90%	0.09%	0.02%
2009	base	13.11%	3.49%	1.39%	3.62%	0.70%	0.20%	0.90%	0.09%	0.02%
2010	60 (08-10)	9.24%	1.48%	0.37%	2.44%	0.34%	0.10%	0.54%	0.03%	0.00%
2010	65 (08-10)	12.12%	2.48%	0.73%	3.28%	0.55%	0.15%	0.77%	0.06%	0.02%
2010	70 (08-10)	14.67%	3.53%	1.20%	4.06%	0.76%	0.25%	1.05%	0.11%	0.02%
2010	75 (08-10)	17.12%	4.77%	1.80%	4.71%	0.98%	0.34%	1.24%	0.14%	0.03%
2010	base	17.12%	4.77%	1.80%	4.71%	0.98%	0.34%	1.24%	0.14%	0.03%

Study area=Houston

year	scenario	FEV10 ages 5to18	FEV15 ages 5to18	FEV20 ages 5to18	FEV10 ages 19to35	FEV15 ages 19to35	FEV20 ages 19to35	FEV10 ages 36to55	FEV15 ages 36to55	FEV20 ages 36to55
2006	60 (06-08)	8.88%	1.29%	0.29%	2.55%	0.28%	0.04%	0.50%	0.03%	0.00%
2006	65 (06-08)	11.63%	2.27%	0.73%	3.29%	0.41%	0.08%	0.70%	0.04%	0.01%
2006	70 (06-08)	13.82%	3.21%	1.12%	3.92%	0.57%	0.11%	0.89%	0.06%	0.01%
2006	75 (06-08)	16.15%	4.20%	1.58%	4.63%	0.77%	0.19%	1.15%	0.09%	0.01%
2006	base	26.83%	10.40%	5.22%	8.54%	2.21%	0.76%	2.74%	0.35%	0.10%
2007	60 (06-08)	7.40%	0.91%	0.22%	2.20%	0.22%	0.04%	0.46%	0.01%	0.00%
2007	65 (06-08)	9.29%	1.48%	0.38%	2.72%	0.31%	0.07%	0.57%	0.02%	0.00%
2007	70 (06-08)	10.74%	2.02%	0.60%	3.14%	0.40%	0.10%	0.67%	0.04%	0.00%
2007	75 (06-08)	12.31%	2.63%	0.83%	3.61%	0.50%	0.14%	0.79%	0.06%	0.01%
2007	base	19.86%	6.61%	2.85%	5.94%	1.30%	0.42%	1.57%	0.17%	0.03%
2008	60 (06-08)	7.53%	0.94%	0.19%	2.23%	0.24%	0.05%	0.51%	0.02%	0.00%
2008	60 (08-10)	9.14%	1.32%	0.33%	2.68%	0.33%	0.07%	0.62%	0.02%	0.00%
2008	65 (06-08)	9.35%	1.39%	0.35%	2.75%	0.34%	0.07%	0.64%	0.02%	0.00%
2008	65 (08-10)	11.68%	2.32%	0.70%	3.51%	0.49%	0.11%	0.86%	0.05%	0.01%
2008	70 (06-08)	10.61%	1.89%	0.52%	3.17%	0.42%	0.09%	0.76%	0.04%	0.00%
2008	70 (08-10)	13.97%	3.25%	1.09%	4.24%	0.67%	0.19%	1.06%	0.07%	0.01%
2008	75 (06-08)	11.91%	2.41%	0.73%	3.56%	0.51%	0.12%	0.89%	0.05%	0.01%
2008	75 (08-10)	16.18%	4.30%	1.62%	4.99%	0.87%	0.25%	1.30%	0.10%	0.02%
2008	base	18.24%	5.58%	2.30%	5.49%	1.07%	0.34%	1.49%	0.16%	0.03%
2009	60 (08-10)	9.29%	1.50%	0.40%	2.72%	0.37%	0.09%	0.58%	0.03%	0.00%
2009	65 (08-10)	12.02%	2.65%	0.89%	3.52%	0.57%	0.12%	0.83%	0.07%	0.01%
2009	70 (08-10)	14.29%	3.78%	1.44%	4.23%	0.76%	0.20%	1.08%	0.10%	0.01%
2009	75 (08-10)	16.61%	4.99%	2.06%	4.96%	1.03%	0.30%	1.31%	0.16%	0.02%
2009	base	19.11%	6.59%	3.10%	5.41%	1.30%	0.44%	1.55%	0.22%	0.06%
2010	60 (08-10)	9.42%	1.48%	0.37%	2.86%	0.35%	0.06%	0.59%	0.04%	0.00%
2010	65 (08-10)	12.21%	2.46%	0.76%	3.60%	0.50%	0.11%	0.76%	0.06%	0.01%
2010	70 (08-10)	14.54%	3.50%	1.21%	4.32%	0.66%	0.17%	0.97%	0.08%	0.02%
2010	75 (08-10)	16.75%	4.63%	1.76%	4.98%	0.87%	0.24%	1.18%	0.10%	0.02%
2010	base	19.23%	6.09%	2.64%	5.53%	1.14%	0.32%	1.44%	0.15%	0.03%

Study area=LosAngeles

year	scenario	FEV10 ages 5to18	FEV15 ages 5to18	FEV20 ages 5to18	FEV10 ages 19to35	FEV15 ages 19to35	FEV20 ages 19to35	FEV10 ages 36to55	FEV15 ages 36to55	FEV20 ages 36to55
2006	60 (06-08)	11.20%	1.70%	0.37%	4.95%	0.59%	0.15%	1.02%	0.05%	0.01%
2006	65 (06-08)	13.55%	2.41%	0.70%	6.11%	0.83%	0.24%	1.32%	0.08%	0.01%
2006	70 (06-08)	15.69%	3.21%	1.06%	7.31%	1.14%	0.33%	1.65%	0.12%	0.03%
2006	75 (06-08)	18.15%	4.22%	1.49%	8.56%	1.50%	0.43%	2.06%	0.19%	0.04%
2006	base	28.03%	11.04%	5.60%	12.69%	3.46%	1.39%	4.12%	0.76%	0.25%
2007	60 (06-08)	10.92%	1.60%	0.37%	4.89%	0.60%	0.12%	0.97%	0.05%	0.01%
2007	65 (06-08)	13.12%	2.29%	0.64%	5.99%	0.85%	0.20%	1.29%	0.08%	0.01%
2007	70 (06-08)	15.08%	3.06%	0.96%	7.03%	1.10%	0.30%	1.63%	0.10%	0.02%
2007	75 (06-08)	17.45%	3.99%	1.40%	8.31%	1.41%	0.44%	2.03%	0.15%	0.03%
2007	base	25.15%	9.32%	4.52%	11.00%	2.80%	1.05%	3.40%	0.52%	0.16%
2008	60 (06-08)	11.60%	1.85%	0.46%	5.10%	0.63%	0.14%	1.08%	0.06%	0.01%
2008	60 (08-10)	11.87%	1.94%	0.48%	5.20%	0.65%	0.15%	1.11%	0.06%	0.01%
2008	65 (06-08)	14.03%	2.56%	0.74%	6.26%	0.88%	0.21%	1.41%	0.09%	0.02%
2008	65 (08-10)	14.29%	2.66%	0.76%	6.40%	0.91%	0.23%	1.45%	0.10%	0.02%
2008	70 (06-08)	16.25%	3.34%	1.06%	7.49%	1.18%	0.31%	1.79%	0.13%	0.02%
2008	70 (08-10)	16.57%	3.48%	1.12%	7.70%	1.25%	0.32%	1.84%	0.14%	0.02%
2008	75 (06-08)	18.60%	4.34%	1.52%	8.91%	1.60%	0.46%	2.23%	0.21%	0.04%
2008	75 (08-10)	18.70%	4.39%	1.54%	8.95%	1.61%	0.46%	2.25%	0.21%	0.04%
2008	base	27.62%	10.51%	5.26%	12.88%	3.40%	1.31%	4.08%	0.69%	0.18%
2009	60 (08-10)	11.92%	1.79%	0.47%	5.27%	0.65%	0.15%	1.12%	0.06%	0.01%
2009	65 (08-10)	14.25%	2.55%	0.74%	6.40%	0.94%	0.25%	1.47%	0.10%	0.02%
2009	70 (08-10)	16.48%	3.39%	1.06%	7.57%	1.26%	0.34%	1.82%	0.13%	0.02%
2009	75 (08-10)	18.55%	4.24%	1.44%	8.75%	1.60%	0.48%	2.21%	0.18%	0.05%
2009	base	25.91%	9.38%	4.53%	11.51%	2.89%	1.06%	3.56%	0.61%	0.16%
2010	60 (08-10)	10.23%	1.47%	0.34%	4.78%	0.59%	0.13%	0.93%	0.06%	0.01%
2010	65 (08-10)	12.33%	2.07%	0.58%	5.77%	0.84%	0.21%	1.25%	0.09%	0.02%
2010	70 (08-10)	14.33%	2.76%	0.89%	6.91%	1.11%	0.29%	1.65%	0.12%	0.03%
2010	75 (08-10)	16.26%	3.52%	1.18%	8.04%	1.38%	0.42%	1.98%	0.14%	0.04%
2010	base	19.70%	6.34%	2.84%	8.34%	1.76%	0.58%	2.41%	0.35%	0.07%

Study area=NewYork

year	scenario	FEV10 ages 5to18	FEV15 ages 5to18	FEV20 ages 5to18	FEV10 ages 19to35	FEV15 ages 19to35	FEV20 ages 19to35	FEV10 ages 36to55	FEV15 ages 36to55	FEV20 ages 36to55
2006	65 (06-08)	2.43%	0.16%	0.03%	0.66%	0.06%	0.01%	0.12%	0.00%	0.00%
2006	70 (06-08)	9.25%	1.85%	0.58%	2.31%	0.29%	0.07%	0.51%	0.03%	0.00%
2006	75 (06-08)	12.72%	3.20%	1.17%	3.20%	0.49%	0.14%	0.78%	0.06%	0.01%
2006	base	23.21%	8.56%	4.09%	6.63%	1.63%	0.62%	2.05%	0.25%	0.06%
2007	65 (06-08)	2.88%	0.26%	0.06%	0.80%	0.08%	0.01%	0.15%	0.01%	0.00%
2007	70 (06-08)	10.62%	2.13%	0.58%	2.73%	0.37%	0.08%	0.61%	0.04%	0.02%
2007	75 (06-08)	14.16%	3.46%	1.18%	3.79%	0.60%	0.15%	0.88%	0.06%	0.02%
2007	base	23.68%	8.52%	3.92%	6.99%	1.69%	0.57%	2.05%	0.24%	0.05%
2008	65 (06-08)	2.58%	0.19%	0.05%	0.68%	0.06%	0.01%	0.12%	0.01%	0.00%
2008	65 (08-10)	3.85%	0.37%	0.07%	1.03%	0.11%	0.02%	0.20%	0.01%	0.00%
2008	70 (06-08)	9.42%	1.85%	0.53%	2.53%	0.38%	0.09%	0.59%	0.05%	0.01%
2008	70 (08-10)	11.69%	2.60%	0.94%	3.17%	0.52%	0.15%	0.77%	0.06%	0.01%
2008	75 (06-08)	12.74%	2.95%	1.14%	3.44%	0.59%	0.18%	0.85%	0.06%	0.02%
2008	75 (08-10)	16.38%	4.56%	1.88%	4.55%	0.90%	0.31%	1.22%	0.12%	0.02%
2008	base	21.41%	7.49%	3.26%	6.18%	1.51%	0.55%	1.81%	0.23%	0.05%
2009	65 (08-10)	3.31%	0.33%	0.06%	0.93%	0.08%	0.02%	0.17%	0.01%	0.00%
2009	70 (08-10)	8.44%	1.58%	0.43%	2.19%	0.31%	0.06%	0.51%	0.03%	0.00%
2009	75 (08-10)	11.41%	2.69%	0.84%	3.01%	0.48%	0.14%	0.75%	0.07%	0.01%
2009	base	13.83%	3.79%	1.47%	3.53%	0.63%	0.20%	0.94%	0.10%	0.01%
2010	65 (08-10)	4.36%	0.49%	0.11%	1.20%	0.12%	0.02%	0.24%	0.01%	0.00%
2010	70 (08-10)	13.62%	3.01%	1.06%	3.59%	0.57%	0.14%	0.91%	0.07%	0.01%
2010	75 (08-10)	18.63%	5.22%	2.08%	5.19%	1.03%	0.30%	1.45%	0.12%	0.03%
2010	base	23.60%	7.86%	3.53%	6.81%	1.60%	0.53%	2.10%	0.24%	0.05%

Study area=Philadelphia

year	scenario	FEV10 ages 5to18	FEV15 ages 5to18	FEV20 ages 5to18	FEV10 ages 19to35	FEV15 ages 19to35	FEV20 ages 19to35	FEV10 ages 36to55	FEV15 ages 36to55	FEV20 ages 36to55
2006	60 (06-08)	9.21%	1.51%	0.35%	2.19%	0.24%	0.05%	0.49%	0.02%	0.01%
2006	65 (06-08)	11.46%	2.15%	0.62%	2.80%	0.36%	0.08%	0.66%	0.03%	0.01%
2006	70 (06-08)	13.90%	3.14%	1.06%	3.50%	0.50%	0.12%	0.92%	0.05%	0.01%
2006	75 (06-08)	16.35%	4.23%	1.59%	4.40%	0.71%	0.20%	1.17%	0.07%	0.01%
2006	base	25.08%	8.82%	4.09%	7.70%	1.77%	0.57%	2.48%	0.26%	0.06%
2007	60 (06-08)	10.39%	1.82%	0.46%	2.52%	0.29%	0.05%	0.58%	0.03%	0.00%
2007	65 (06-08)	12.82%	2.70%	0.84%	3.24%	0.44%	0.11%	0.78%	0.05%	0.01%
2007	70 (06-08)	15.69%	3.89%	1.40%	4.20%	0.67%	0.17%	1.08%	0.09%	0.01%
2007	75 (06-08)	18.41%	5.17%	2.03%	5.13%	0.86%	0.27%	1.41%	0.12%	0.03%
2007	base	27.22%	10.21%	4.97%	8.76%	2.22%	0.79%	2.88%	0.38%	0.10%
2008	60 (06-08)	8.19%	1.22%	0.24%	1.92%	0.21%	0.04%	0.46%	0.02%	0.01%
2008	60 (08-10)	10.17%	1.84%	0.48%	2.41%	0.30%	0.05%	0.60%	0.04%	0.01%
2008	65 (06-08)	10.17%	1.84%	0.48%	2.41%	0.30%	0.05%	0.60%	0.04%	0.01%
2008	65 (08-10)	12.60%	2.75%	0.88%	3.14%	0.44%	0.12%	0.80%	0.06%	0.01%
2008	70 (06-08)	12.60%	2.75%	0.88%	3.14%	0.44%	0.12%	0.80%	0.06%	0.01%
2008	70 (08-10)	15.57%	4.01%	1.50%	4.15%	0.70%	0.19%	1.12%	0.10%	0.02%
2008	75 (06-08)	14.90%	3.72%	1.36%	3.96%	0.64%	0.17%	1.04%	0.10%	0.02%
2008	75 (08-10)	18.64%	5.38%	2.23%	5.27%	0.99%	0.29%	1.52%	0.16%	0.03%
2008	base	23.84%	8.19%	3.78%	7.30%	1.64%	0.56%	2.25%	0.31%	0.07%
2009	60 (08-10)	7.60%	0.99%	0.21%	1.82%	0.20%	0.03%	0.40%	0.02%	0.00%
2009	65 (08-10)	9.04%	1.44%	0.34%	2.19%	0.26%	0.05%	0.51%	0.03%	0.00%
2009	70 (08-10)	10.75%	2.01%	0.57%	2.73%	0.35%	0.08%	0.63%	0.04%	0.01%
2009	75 (08-10)	12.51%	2.73%	0.86%	3.25%	0.48%	0.12%	0.78%	0.06%	0.01%
2009	base	15.31%	4.06%	1.54%	4.03%	0.69%	0.19%	1.03%	0.10%	0.02%
2010	60 (08-10)	11.26%	1.99%	0.55%	2.65%	0.31%	0.04%	0.65%	0.04%	0.01%
2010	65 (08-10)	13.74%	2.92%	0.95%	3.43%	0.45%	0.10%	0.89%	0.06%	0.01%
2010	70 (08-10)	16.88%	4.16%	1.50%	4.46%	0.73%	0.18%	1.20%	0.11%	0.02%
2010	75 (08-10)	19.76%	5.51%	2.24%	5.59%	1.04%	0.29%	1.61%	0.15%	0.03%
2010	base	24.84%	8.23%	3.70%	7.45%	1.63%	0.57%	2.37%	0.29%	0.07%

Study area=Sacramento

year	scenario	FEV10 ages 5to18	FEV15 ages 5to18	FEV20 ages 5to18	FEV10 ages 19to35	FEV15 ages 19to35	FEV20 ages 19to35	FEV10 ages 36to55	FEV15 ages 36to55	FEV20 ages 36to55
2006	60 (06-08)	9.76%	1.64%	0.42%	2.40%	0.25%	0.03%	0.50%	0.03%	0.00%
2006	65 (06-08)	12.85%	2.72%	0.83%	3.45%	0.40%	0.10%	0.78%	0.06%	0.01%
2006	70 (06-08)	15.25%	3.63%	1.28%	4.20%	0.59%	0.14%	1.03%	0.07%	0.01%
2006	75 (06-08)	17.92%	4.75%	1.77%	5.15%	0.85%	0.21%	1.31%	0.12%	0.02%
2006	base	29.22%	10.91%	5.23%	9.95%	2.51%	0.91%	3.07%	0.52%	0.12%
2007	60 (06-08)	7.39%	0.97%	0.25%	1.80%	0.14%	0.02%	0.37%	0.02%	0.00%
2007	65 (06-08)	9.59%	1.62%	0.48%	2.35%	0.23%	0.03%	0.49%	0.03%	0.00%
2007	70 (06-08)	11.20%	2.17%	0.69%	2.78%	0.30%	0.07%	0.61%	0.04%	0.00%
2007	75 (06-08)	13.04%	2.87%	0.96%	3.35%	0.46%	0.10%	0.77%	0.07%	0.01%
2007	base	21.38%	6.95%	3.03%	6.22%	1.35%	0.41%	1.76%	0.22%	0.05%
2008	60 (06-08)	9.41%	1.61%	0.49%	2.38%	0.25%	0.06%	0.49%	0.02%	0.00%
2008	60 (08-10)	9.41%	1.61%	0.49%	2.38%	0.25%	0.06%	0.49%	0.02%	0.00%
2008	65 (06-08)	12.30%	2.65%	0.86%	3.26%	0.43%	0.11%	0.72%	0.05%	0.00%
2008	65 (08-10)	11.97%	2.49%	0.81%	3.15%	0.40%	0.10%	0.69%	0.05%	0.00%
2008	70 (06-08)	14.43%	3.44%	1.22%	3.99%	0.60%	0.14%	0.92%	0.09%	0.00%
2008	70 (08-10)	14.09%	3.33%	1.16%	3.87%	0.57%	0.13%	0.87%	0.07%	0.00%
2008	75 (06-08)	16.79%	4.45%	1.77%	4.85%	0.86%	0.20%	1.21%	0.13%	0.01%
2008	75 (08-10)	16.43%	4.29%	1.67%	4.75%	0.82%	0.19%	1.16%	0.12%	0.01%
2008	base	27.44%	10.39%	4.96%	9.51%	2.27%	0.86%	2.92%	0.48%	0.13%
2009	60 (08-10)	8.61%	1.37%	0.36%	2.18%	0.23%	0.04%	0.43%	0.03%	0.00%
2009	65 (08-10)	10.98%	2.15%	0.68%	2.87%	0.38%	0.09%	0.61%	0.05%	0.01%
2009	70 (08-10)	13.02%	2.88%	1.01%	3.56%	0.52%	0.13%	0.80%	0.07%	0.01%
2009	75 (08-10)	15.38%	3.81%	1.41%	4.33%	0.75%	0.19%	1.04%	0.10%	0.02%
2009	base	25.55%	8.97%	4.30%	8.32%	2.14%	0.78%	2.44%	0.36%	0.10%
2010	60 (08-10)	6.56%	0.77%	0.19%	1.66%	0.13%	0.02%	0.32%	0.02%	0.00%
2010	65 (08-10)	8.30%	1.29%	0.36%	2.10%	0.18%	0.03%	0.42%	0.02%	0.00%
2010	70 (08-10)	9.81%	1.75%	0.53%	2.47%	0.24%	0.04%	0.54%	0.04%	0.00%
2010	75 (08-10)	11.50%	2.35%	0.77%	3.02%	0.33%	0.07%	0.68%	0.05%	0.01%
2010	base	19.22%	5.96%	2.54%	5.44%	1.02%	0.26%	1.50%	0.17%	0.04%

Study area=StLouis

year	scenario	FEV10 ages 5to18	FEV15 ages 5to18	FEV20 ages 5to18	FEV10 ages 19to35	FEV15 ages 19to35	FEV20 ages 19to35	FEV10 ages 36to55	FEV15 ages 36to55	FEV20 ages 36to55
2006	60 (06-08)	8.82%	1.39%	0.32%	2.33%	0.29%	0.07%	0.49%	0.03%	0.01%
2006	65 (06-08)	11.80%	2.38%	0.71%	3.26%	0.49%	0.13%	0.75%	0.06%	0.01%
2006	70 (06-08)	15.20%	3.65%	1.25%	4.43%	0.74%	0.22%	1.08%	0.11%	0.02%
2006	75 (06-08)	18.64%	5.11%	2.00%	5.73%	1.12%	0.36%	1.49%	0.15%	0.04%
2006	base	24.21%	8.11%	3.58%	7.99%	1.87%	0.64%	2.35%	0.28%	0.07%
2007	60 (06-08)	10.00%	1.92%	0.53%	2.53%	0.31%	0.07%	0.57%	0.04%	0.00%
2007	65 (06-08)	13.43%	3.16%	1.06%	3.71%	0.54%	0.13%	0.84%	0.07%	0.02%
2007	70 (06-08)	17.08%	4.66%	1.82%	5.13%	0.90%	0.25%	1.28%	0.12%	0.03%
2007	75 (06-08)	20.80%	6.44%	2.77%	6.69%	1.36%	0.43%	1.78%	0.20%	0.06%
2007	base	27.07%	9.70%	4.64%	9.20%	2.30%	0.83%	2.79%	0.38%	0.10%
2008	60 (06-08)	5.72%	0.69%	0.12%	1.56%	0.18%	0.04%	0.31%	0.01%	0.00%
2008	60 (08-10)	8.23%	1.31%	0.31%	2.22%	0.27%	0.08%	0.49%	0.02%	0.00%
2008	65 (06-08)	7.62%	1.14%	0.24%	2.02%	0.24%	0.06%	0.45%	0.02%	0.00%
2008	65 (08-10)	10.82%	2.08%	0.65%	3.04%	0.42%	0.13%	0.69%	0.05%	0.01%
2008	70 (06-08)	9.61%	1.67%	0.49%	2.69%	0.34%	0.10%	0.60%	0.03%	0.01%
2008	70 (08-10)	13.20%	2.96%	1.04%	3.79%	0.60%	0.17%	0.91%	0.08%	0.02%
2008	75 (06-08)	11.73%	2.46%	0.78%	3.34%	0.49%	0.14%	0.77%	0.06%	0.01%
2008	75 (08-10)	14.64%	3.57%	1.29%	4.26%	0.71%	0.19%	1.07%	0.11%	0.02%
2008	base	15.08%	3.80%	1.41%	4.38%	0.76%	0.20%	1.12%	0.11%	0.02%
2009	60 (08-10)	7.25%	1.11%	0.27%	1.91%	0.20%	0.05%	0.39%	0.02%	0.00%
2009	65 (08-10)	9.64%	1.81%	0.54%	2.58%	0.34%	0.08%	0.54%	0.04%	0.01%
2009	70 (08-10)	11.72%	2.62%	0.88%	3.29%	0.47%	0.13%	0.73%	0.06%	0.01%
2009	75 (08-10)	13.39%	3.29%	1.19%	3.76%	0.61%	0.18%	0.87%	0.09%	0.01%
2009	base	13.86%	3.53%	1.32%	3.89%	0.65%	0.22%	0.93%	0.09%	0.01%
2010	60 (08-10)	9.91%	1.79%	0.51%	2.75%	0.41%	0.10%	0.67%	0.05%	0.00%
2010	65 (08-10)	13.17%	3.01%	0.99%	3.92%	0.66%	0.18%	1.00%	0.09%	0.01%
2010	70 (08-10)	16.14%	4.19%	1.54%	4.95%	0.94%	0.29%	1.31%	0.14%	0.03%
2010	75 (08-10)	18.44%	5.26%	2.09%	5.68%	1.18%	0.39%	1.59%	0.18%	0.05%
2010	base	19.07%	5.59%	2.27%	5.94%	1.27%	0.44%	1.70%	0.19%	0.05%

Study area=WashingtonDC

year	scenario	FEV10 ages 5to18	FEV15 ages 5to18	FEV20 ages 5to18	FEV10 ages 19to35	FEV15 ages 19to35	FEV20 ages 19to35	FEV10 ages 36to55	FEV15 ages 36to55	FEV20 ages 36to55
2006	60 (06-08)	7.73%	1.19%	0.30%	2.01%	0.26%	0.07%	0.41%	0.03%	0.01%
2006	65 (06-08)	10.45%	1.98%	0.63%	2.81%	0.40%	0.11%	0.56%	0.05%	0.01%
2006	70 (06-08)	13.16%	2.94%	1.00%	3.69%	0.57%	0.17%	0.79%	0.08%	0.02%
2006	75 (06-08)	15.89%	4.00%	1.49%	4.51%	0.79%	0.25%	1.04%	0.10%	0.03%
2006	base	25.95%	9.34%	4.41%	8.38%	2.03%	0.73%	2.38%	0.30%	0.08%
2007	60 (06-08)	8.97%	1.38%	0.32%	2.39%	0.29%	0.06%	0.50%	0.04%	0.01%
2007	65 (06-08)	11.65%	2.23%	0.61%	3.20%	0.45%	0.09%	0.72%	0.06%	0.01%
2007	70 (06-08)	14.33%	3.28%	1.04%	4.09%	0.67%	0.17%	0.97%	0.09%	0.02%
2007	75 (06-08)	16.89%	4.34%	1.51%	4.98%	0.91%	0.26%	1.22%	0.12%	0.03%
2007	base	26.33%	9.23%	4.23%	8.64%	2.08%	0.74%	2.49%	0.33%	0.10%
2008	60 (06-08)	6.99%	0.97%	0.21%	1.85%	0.23%	0.05%	0.34%	0.02%	0.00%
2008	60 (08-10)	7.82%	1.19%	0.29%	2.07%	0.26%	0.06%	0.38%	0.03%	0.00%
2008	65 (06-08)	9.07%	1.54%	0.41%	2.42%	0.32%	0.07%	0.45%	0.03%	0.00%
2008	65 (08-10)	11.12%	2.24%	0.70%	3.03%	0.43%	0.12%	0.63%	0.05%	0.01%
2008	70 (06-08)	11.12%	2.24%	0.70%	3.03%	0.43%	0.12%	0.63%	0.05%	0.01%
2008	70 (08-10)	14.05%	3.37%	1.24%	3.92%	0.67%	0.20%	0.94%	0.07%	0.02%
2008	75 (06-08)	13.22%	3.05%	1.06%	3.66%	0.60%	0.17%	0.84%	0.06%	0.02%
2008	75 (08-10)	17.96%	5.11%	2.12%	5.23%	1.03%	0.33%	1.35%	0.13%	0.03%
2008	base	21.46%	6.92%	3.08%	6.29%	1.37%	0.45%	1.72%	0.17%	0.04%
2009	60 (08-10)	6.15%	0.79%	0.15%	1.63%	0.18%	0.04%	0.32%	0.02%	0.01%
2009	65 (08-10)	8.30%	1.31%	0.34%	2.14%	0.26%	0.07%	0.44%	0.03%	0.01%
2009	70 (08-10)	10.12%	1.83%	0.57%	2.64%	0.35%	0.09%	0.59%	0.05%	0.01%
2009	75 (08-10)	12.33%	2.73%	0.88%	3.30%	0.47%	0.13%	0.78%	0.07%	0.01%
2009	base	14.05%	3.47%	1.17%	3.61%	0.58%	0.15%	0.91%	0.09%	0.01%
2010	60 (08-10)	9.68%	1.59%	0.39%	2.67%	0.35%	0.08%	0.59%	0.04%	0.01%
2010	65 (08-10)	13.39%	2.87%	0.92%	3.93%	0.63%	0.16%	0.96%	0.07%	0.02%
2010	70 (08-10)	16.80%	4.20%	1.58%	5.07%	0.93%	0.29%	1.29%	0.12%	0.03%
2010	75 (08-10)	21.14%	6.22%	2.54%	6.65%	1.41%	0.46%	1.82%	0.19%	0.05%
2010	base	24.63%	8.24%	3.57%	7.87%	1.83%	0.66%	2.25%	0.28%	0.07%

Counts of person-days with lung function decrement > 10,15,20%, study area=Atlanta

year	scenario	FEV10 ages 5to18	FEV15 ages 5to18	FEV20 ages 5to18	FEV10 ages 19to35	FEV15 ages 19to35	FEV20 ages 19to35	FEV10 ages 36to55	FEV15 ages 36to55	FEV20 ages 36to55
2006	60 (06-08)	311,580	39,934	8,395	56,647	5,199	982	12,188	443	19
2006	65 (06-08)	458,918	73,149	18,369	84,278	9,281	1,887	18,446	963	58
2006	70 (06-08)	602,077	110,599	31,597	111,119	15,115	2,965	25,782	1,425	135
2006	75 (06-08)	806,774	171,964	53,894	151,169	24,300	5,295	36,391	2,753	366
2006	base	1,664,477	488,705	194,126	331,451	70,935	23,549	93,540	11,014	2,368
2007	60 (06-08)	313,698	41,918	8,780	55,550	5,988	1,078	9,916	193	39
2007	65 (06-08)	462,384	74,843	19,081	81,813	10,186	2,137	16,078	732	58
2007	70 (06-08)	608,547	111,119	31,212	107,673	14,595	3,466	23,452	1,348	96
2007	75 (06-08)	811,049	170,539	53,990	147,357	23,452	5,699	33,715	2,137	231
2007	base	1,686,832	490,111	191,623	329,545	71,685	22,721	86,954	9,493	1,694
2008	60 (06-08)	190,256	18,388	3,331	37,065	3,716	616	7,413	347	19
2008	60 (08-10)	301,992	39,376	8,126	55,646	6,701	1,290	11,418	693	39
2008	65 (06-08)	273,745	34,312	6,489	51,564	5,911	1,098	10,513	616	19
2008	65 (08-10)	414,882	64,561	16,174	76,287	9,916	2,195	15,962	1,155	154
2008	70 (06-08)	357,330	51,333	11,514	65,543	8,010	1,771	13,671	905	96
2008	70 (08-10)	577,700	107,326	30,827	108,231	15,385	3,909	23,144	1,771	308
2008	75 (06-08)	473,609	79,618	21,315	87,667	11,880	2,927	18,485	1,348	212
2008	75 (08-10)	794,740	171,310	55,377	146,702	23,895	6,104	34,177	3,100	539
2008	base	990,773	240,993	84,047	182,439	32,329	9,146	45,268	4,467	809
2009	60 (08-10)	192,567	22,066	4,409	37,065	4,043	732	6,701	347	58
2009	65 (08-10)	261,133	36,526	8,491	49,388	6,007	1,175	9,570	539	58
2009	70 (08-10)	360,372	60,133	16,058	67,064	9,107	2,022	13,806	751	96
2009	75 (08-10)	491,901	96,466	29,190	90,209	13,209	3,350	19,794	1,348	231
2009	base	602,905	131,452	42,957	108,269	17,811	4,756	25,070	2,080	347
2010	60 (08-10)	304,726	38,952	7,702	51,892	5,083	847	9,839	443	135
2010	65 (08-10)	413,284	61,269	13,767	69,086	7,529	1,444	13,786	578	154
2010	70 (08-10)	562,778	98,796	25,532	94,252	11,264	2,619	20,391	1,059	193
2010	75 (08-10)	745,544	149,667	44,324	124,270	16,540	4,101	27,785	1,906	250
2010	base	902,605	199,036	64,407	149,841	22,528	5,796	34,639	2,580	327

Counts of person-days with lung function decrement > 10,15,20%, study area=Baltimore

year	scenario	FEV10 ages 5to18	FEV15 ages 5to18	FEV20 ages 5to18	FEV10 ages 19to35	FEV15 ages 19to35	FEV20 ages 19to35	FEV10 ages 36to55	FEV15 ages 36to55	FEV20 ages 36to55
2006	60 (06-08)	161,638	19,552	4,131	22,600	3,060	707	5,313	287	55
2006	65 (06-08)	239,557	38,463	8,969	32,796	4,684	1,270	8,285	574	99
2006	70 (06-08)	315,268	60,113	16,160	43,511	6,606	1,977	11,499	917	155
2006	75 (06-08)	408,143	87,585	26,356	56,921	9,356	2,806	15,697	1,314	199
2006	base	718,727	203,624	78,008	107,479	21,849	7,003	32,895	3,634	729
2007	60 (06-08)	158,733	19,408	3,711	20,513	2,342	641	4,838	243	11
2007	65 (06-08)	226,136	34,596	8,119	29,471	3,601	1,016	7,070	365	33
2007	70 (06-08)	292,071	51,066	13,123	37,999	4,971	1,447	9,577	574	77
2007	75 (06-08)	365,218	73,214	21,186	48,040	7,047	1,944	13,012	851	122
2007	base	584,086	153,011	54,678	79,190	14,824	4,507	24,511	2,408	376
2008	60 (06-08)	108,617	11,134	1,922	15,420	1,679	320	3,491	177	33
2008	60 (08-10)	118,912	12,935	2,397	16,901	1,900	376	3,800	221	44
2008	65 (06-08)	151,962	19,596	4,717	21,065	2,452	519	4,982	331	55
2008	65 (08-10)	157,960	20,844	5,059	21,860	2,530	563	5,114	331	55
2008	70 (06-08)	195,064	29,946	7,456	27,052	3,325	862	6,462	497	88
2008	70 (08-10)	207,391	33,083	8,550	28,841	3,623	972	7,003	608	99
2008	75 (06-08)	244,771	43,975	11,852	34,353	4,728	1,193	8,483	773	155
2008	75 (08-10)	256,811	47,752	13,145	36,055	5,070	1,326	9,157	862	155
2008	base	402,731	97,780	33,856	57,186	10,019	2,938	16,669	1,569	376
2009	60 (08-10)	93,947	8,152	1,160	14,437	1,624	353	3,281	99	11
2009	65 (08-10)	121,187	13,211	2,132	18,171	2,121	464	4,164	166	22
2009	70 (08-10)	154,635	20,468	3,667	22,545	2,828	685	5,379	320	44
2009	75 (08-10)	185,774	27,450	5,877	26,776	3,579	817	6,440	453	66
2009	base	268,399	52,215	14,194	36,762	5,479	1,535	9,356	696	133
2010	60 (08-10)	190,800	24,655	4,639	25,307	2,861	751	5,379	320	11
2010	65 (08-10)	256,845	40,053	9,290	34,066	4,220	1,160	7,799	519	44
2010	70 (08-10)	338,663	61,549	16,293	45,764	6,517	1,767	10,869	862	66
2010	75 (08-10)	420,615	86,392	25,638	57,760	8,848	2,585	14,581	1,071	155
2010	base	639,151	167,868	59,428	89,827	17,033	5,523	26,643	2,684	530

Counts of person-days with lung function decrement > 10,15,20%, study area=Boston

year	scenario	FEV10 ages 5to18	FEV15 ages 5to18	FEV20 ages 5to18	FEV10 ages 19to35	FEV15 ages 19to35	FEV20 ages 19to35	FEV10 ages 36to55	FEV15 ages 36to55	FEV20 ages 36to55
2006	60 (06-08)	166,848	18,420	3,582	31,056	3,181	734	7,030	334	0
2006	65 (06-08)	233,143	33,036	7,564	42,891	4,805	1,179	10,901	556	22
2006	70 (06-08)	331,027	59,265	15,750	59,999	8,498	2,069	17,219	1,112	200
2006	75 (06-08)	399,858	82,201	24,093	72,079	11,546	3,003	22,180	1,802	289
2006	base	533,425	130,431	45,650	95,148	17,219	5,117	31,301	3,203	645
2007	60 (06-08)	221,397	30,255	6,518	36,195	3,404	690	9,099	556	133
2007	65 (06-08)	320,193	56,261	14,571	52,702	6,118	1,179	14,060	1,001	222
2007	70 (06-08)	464,840	102,089	30,878	77,529	10,990	2,736	23,759	1,958	423
2007	75 (06-08)	573,425	140,553	46,918	98,085	15,817	4,271	30,722	2,937	623
2007	base	766,769	215,635	82,846	131,632	24,760	7,453	46,740	5,206	1,201
2008	60 (06-08)	158,973	17,063	2,759	30,811	3,203	578	7,052	311	44
2008	60 (08-10)	192,921	23,759	4,316	35,750	3,849	756	9,054	378	44
2008	65 (06-08)	221,397	30,010	6,585	40,733	4,650	979	10,612	445	44
2008	65 (08-10)	280,950	44,604	10,656	50,499	6,563	1,379	13,949	623	67
2008	70 (06-08)	311,673	52,680	13,237	55,549	7,430	1,691	15,795	868	67
2008	70 (08-10)	379,124	71,411	19,777	67,807	9,811	2,425	20,022	1,290	133
2008	75 (06-08)	379,146	71,411	19,777	67,807	9,811	2,425	20,022	1,290	133
2008	75 (08-10)	458,855	99,731	31,412	80,866	13,014	3,515	24,916	1,735	311
2008	base	487,019	110,943	36,106	85,360	14,371	3,915	27,341	2,024	423
2009	60 (08-10)	161,443	21,379	4,939	31,568	3,493	823	6,474	245	67
2009	65 (08-10)	218,482	35,127	8,943	41,868	4,939	1,313	9,588	556	89
2009	70 (08-10)	276,256	50,833	14,994	51,656	6,807	1,935	13,103	845	89
2009	75 (08-10)	305,622	60,488	18,821	53,614	7,942	2,269	14,861	823	111
2009	base	312,540	64,137	19,977	54,170	8,142	2,247	15,372	868	89
2010	60 (08-10)	209,005	24,582	4,338	36,062	3,804	645	8,921	534	67
2010	65 (08-10)	297,613	44,181	9,633	49,098	5,495	1,023	14,015	934	89
2010	70 (08-10)	392,516	68,831	17,575	64,159	8,164	1,757	18,776	1,268	133
2010	75 (08-10)	459,945	89,942	26,206	73,013	9,344	2,113	21,979	1,424	178
2010	base	480,768	97,929	29,365	75,326	9,877	2,291	23,159	1,646	200

Counts of person-days with lung function decrement > 10,15,20%, study area=Chicago

year	scenario	FEV10 ages 5to18	FEV15 ages 5to18	FEV20 ages 5to18	FEV10 ages 19to35	FEV15 ages 19to35	FEV20 ages 19to35	FEV10 ages 36to55	FEV15 ages 36to55	FEV20 ages 36to55
2006	60 (06-08)	473,892	63,467	13,812	87,084	11,058	2,796	16,440	835	83
2006	65 (06-08)	630,952	101,021	24,994	110,701	15,898	3,672	22,574	1,460	209
2006	70 (06-08)	785,926	144,083	40,892	133,484	20,655	5,591	29,125	2,253	417
2006	75 (06-08)	890,869	187,562	57,667	145,126	24,494	6,676	35,510	3,380	793
2006	base	881,897	192,695	59,836	140,494	23,909	6,551	35,092	3,380	709
2007	60 (06-08)	715,699	106,445	23,325	110,660	13,645	3,380	22,157	793	83
2007	65 (06-08)	1,003,615	183,181	48,236	153,763	21,573	5,675	35,259	1,753	292
2007	70 (06-08)	1,301,544	274,897	81,952	200,957	30,878	8,512	48,904	3,296	501
2007	75 (06-08)	1,560,543	372,579	125,640	236,675	39,599	12,226	63,550	4,882	751
2007	base	1,579,612	390,897	134,945	235,673	40,642	12,518	64,760	5,341	876
2008	60 (06-08)	405,752	40,642	7,803	70,393	8,178	1,878	12,351	626	42
2008	60 (08-10)	498,094	56,665	11,183	81,868	9,347	2,462	15,439	835	42
2008	65 (06-08)	532,852	63,675	12,894	86,583	10,307	2,712	16,941	793	42
2008	65 (08-10)	626,279	86,500	18,109	97,975	11,725	3,296	20,780	1,127	125
2008	70 (06-08)	649,312	92,425	19,945	100,729	12,518	3,463	21,698	1,168	125
2008	70 (08-10)	714,155	114,874	28,291	104,860	13,645	3,338	24,368	1,377	209
2008	75 (06-08)	705,351	114,248	28,833	102,940	13,186	3,130	24,035	1,377	209
2008	75 (08-10)	677,853	111,870	28,666	96,264	11,976	2,921	23,576	1,335	125
2008	base	677,853	111,870	28,666	96,264	11,976	2,921	23,576	1,335	125
2009	60 (08-10)	454,197	61,172	14,855	88,544	10,766	2,837	16,566	960	125
2009	65 (08-10)	570,782	90,339	23,909	107,155	13,978	3,881	22,074	1,252	209
2009	70 (08-10)	648,561	118,463	32,213	116,251	16,524	4,340	25,871	1,460	250
2009	75 (08-10)	616,389	116,126	32,338	103,357	15,272	3,881	24,535	1,502	250
2009	base	616,389	116,126	32,338	103,357	15,272	3,881	24,535	1,502	250
2010	60 (08-10)	730,429	104,776	23,117	129,979	18,068	4,465	25,245	1,210	250
2010	65 (08-10)	935,600	162,860	40,308	162,401	24,619	6,176	34,466	1,669	292
2010	70 (08-10)	1,093,077	218,357	61,464	184,808	30,628	8,053	42,520	2,629	584
2010	75 (08-10)	1,064,160	222,780	65,511	172,374	29,459	7,636	40,892	2,754	668
2010	base	1,064,160	222,780	65,511	172,374	29,459	7,636	40,892	2,754	668

Counts of person-days with lung function decrement > 10,15,20%, study area=Cleveland

year	scenario	FEV10 ages 5to18	FEV15 ages 5to18	FEV20 ages 5to18	FEV10 ages 19to35	FEV15 ages 19to35	FEV20 ages 19to35	FEV10 ages 36to55	FEV15 ages 36to55	FEV20 ages 36to55
2006	60 (06-08)	59,108	3,918	431	10,542	673	67	2,827	148	0
2006	65 (06-08)	125,864	14,945	2,572	21,193	2,302	269	5,992	296	81
2006	70 (06-08)	191,704	28,948	6,584	32,220	4,174	808	9,385	741	162
2006	75 (06-08)	264,384	48,364	13,262	44,284	6,880	1,535	13,249	1,185	229
2006	base	373,363	84,407	27,857	63,201	11,566	3,124	20,870	2,222	539
2007	60 (06-08)	79,062	7,082	1,319	11,135	929	162	3,110	135	0
2007	65 (06-08)	179,344	25,097	6,113	24,774	2,652	646	6,665	404	67
2007	70 (06-08)	279,194	48,687	13,626	40,110	4,928	1,225	11,768	875	148
2007	75 (06-08)	389,682	80,099	25,717	58,475	8,429	2,275	18,338	1,575	188
2007	base	544,480	138,749	50,908	85,606	15,215	4,268	30,281	2,881	525
2008	60 (06-08)	68,923	5,036	539	11,566	902	175	2,760	94	0
2008	60 (08-10)	68,923	5,036	539	11,566	902	175	2,760	94	0
2008	65 (06-08)	151,405	17,800	3,635	23,132	2,329	458	6,517	256	27
2008	65 (08-10)	127,627	13,774	2,343	19,604	1,818	337	5,332	202	27
2008	70 (06-08)	230,306	35,007	8,725	35,249	4,268	875	10,192	525	94
2008	70 (08-10)	210,984	29,971	7,298	31,816	3,676	794	9,115	404	67
2008	75 (06-08)	315,844	58,556	16,521	48,902	6,597	1,602	14,797	1,077	135
2008	75 (08-10)	303,241	55,338	15,309	46,923	6,113	1,468	13,962	969	108
2008	base	429,899	97,091	30,416	66,177	10,879	2,733	21,691	1,818	337
2009	60 (08-10)	58,098	5,089	687	10,556	687	81	2,316	108	13
2009	65 (08-10)	92,822	10,596	2,073	15,780	1,454	202	3,810	215	27
2009	70 (08-10)	139,597	19,954	4,537	23,226	2,612	404	6,301	417	27
2009	75 (08-10)	189,186	31,520	8,429	31,385	4,160	902	9,142	673	40
2009	base	239,192	46,667	14,097	38,777	5,763	1,495	11,593	942	148
2010	60 (08-10)	76,679	5,413	646	12,333	902	94	3,137	81	0
2010	65 (08-10)	137,779	13,989	2,329	20,291	1,952	296	5,938	202	13
2010	70 (08-10)	225,970	31,991	6,557	32,839	3,797	902	9,815	539	67
2010	75 (08-10)	323,949	57,115	14,636	48,188	6,422	1,481	14,663	983	135
2010	base	455,037	98,154	30,079	67,671	10,906	2,881	21,502	1,737	350

Counts of person-days with lung function decrement > 10,15,20%, study area=Dallas

year	scenario	FEV10 ages 5to18	FEV15 ages 5to18	FEV20 ages 5to18	FEV10 ages 19to35	FEV15 ages 19to35	FEV20 ages 19to35	FEV10 ages 36to55	FEV15 ages 36to55	FEV20 ages 36to55
2006	60 (06-08)	611,002	93,592	21,707	93,991	9,679	1,621	22,317	1,081	94
2006	65 (06-08)	833,072	151,053	40,242	130,756	15,904	3,242	31,667	2,091	258
2006	70 (06-08)	1,112,297	230,621	69,184	181,053	25,019	6,155	46,538	3,218	540
2006	75 (06-08)	1,362,604	312,044	101,274	228,812	34,862	8,880	60,280	5,098	799
2006	base	2,225,534	653,288	257,026	397,249	81,212	24,620	116,215	13,343	2,443
2007	60 (06-08)	319,068	38,245	7,142	52,035	5,051	1,034	10,783	446	47
2007	65 (06-08)	427,460	61,643	14,072	70,147	7,353	1,644	15,246	916	47
2007	70 (06-08)	562,092	96,834	25,301	92,182	11,417	2,537	21,213	1,433	211
2007	75 (06-08)	680,727	131,884	38,527	113,325	15,082	3,688	27,133	2,161	305
2007	base	1,054,648	271,473	98,243	182,368	34,392	10,008	49,145	5,403	1,034
2008	60 (06-08)	330,908	36,084	5,920	51,588	4,698	681	11,535	305	23
2008	60 (08-10)	330,908	36,084	5,920	51,588	4,698	681	11,535	305	23
2008	65 (06-08)	444,350	58,166	11,840	68,503	7,071	1,339	15,481	470	47
2008	65 (08-10)	444,350	58,166	11,840	68,503	7,071	1,339	15,481	470	47
2008	70 (06-08)	578,959	89,199	21,331	89,645	10,125	2,349	20,720	1,057	94
2008	70 (08-10)	578,959	89,199	21,331	89,645	10,125	2,349	20,720	1,057	94
2008	75 (06-08)	695,432	121,829	31,150	109,520	13,813	3,359	25,630	1,339	164
2008	75 (08-10)	729,848	130,451	34,204	115,604	14,589	3,547	27,274	1,433	188
2008	base	1,051,171	241,662	78,369	169,095	26,405	7,564	43,554	3,383	423
2009	60 (08-10)	373,381	48,182	9,820	58,072	6,108	1,386	12,897	470	23
2009	65 (08-10)	499,510	76,325	18,324	79,473	9,021	2,255	18,183	916	70
2009	70 (08-10)	660,782	118,306	31,973	106,372	14,142	3,524	25,042	1,621	117
2009	75 (08-10)	842,539	171,421	52,129	139,260	20,626	5,474	34,322	2,490	305
2009	base	1,263,562	332,270	120,044	216,666	40,195	12,827	58,096	6,225	893
2010	60 (08-10)	317,329	34,016	6,084	51,283	4,933	799	10,830	352	23
2010	65 (08-10)	417,945	54,055	11,065	67,798	7,470	1,339	14,870	775	23
2010	70 (08-10)	541,607	83,490	19,193	88,565	10,712	1,997	19,733	1,269	117
2010	75 (08-10)	679,223	120,138	31,526	111,281	15,458	3,336	26,593	1,926	211
2010	base	951,401	217,395	69,701	154,460	26,475	7,470	41,933	3,876	634

Counts of person-days with lung function decrement > 10,15,20%, study area=Denver

year	scenario	FEV10 ages 5to18	FEV15 ages 5to18	FEV20 ages 5to18	FEV10 ages 19to35	FEV15 ages 19to35	FEV20 ages 19to35	FEV10 ages 36to55	FEV15 ages 36to55	FEV20 ages 36to55
2006	60 (06-08)	296,332	41,783	8,969	55,427	6,894	1,405	14,379	880	39
2006	65 (06-08)	399,123	67,691	16,729	74,034	10,072	2,206	20,261	1,274	144
2006	70 (06-08)	504,566	99,521	27,536	93,494	13,958	3,611	25,566	2,101	315
2006	75 (06-08)	611,559	134,398	41,206	113,637	17,977	4,990	32,263	2,955	499
2006	base	777,551	201,419	71,119	145,848	28,166	8,417	44,068	5,134	1,011
2007	60 (06-08)	265,079	33,379	6,237	49,255	5,581	1,103	12,304	709	79
2007	65 (06-08)	355,435	55,427	12,081	64,763	8,233	1,602	16,860	1,300	131
2007	70 (06-08)	441,970	80,074	19,999	79,142	10,610	2,429	21,325	1,891	276
2007	75 (06-08)	527,572	106,494	29,926	93,809	13,775	3,493	25,947	2,456	368
2007	base	647,027	153,228	49,163	112,403	18,830	5,226	32,999	3,729	591
2008	60 (06-08)	292,248	40,116	7,669	58,801	6,894	1,444	14,142	801	105
2008	60 (08-10)	292,248	40,116	7,669	58,801	6,894	1,444	14,142	801	105
2008	65 (06-08)	398,059	64,343	14,221	78,525	10,137	2,127	19,736	1,287	184
2008	65 (08-10)	513,246	96,068	24,700	100,441	13,998	3,598	25,947	1,970	302
2008	70 (06-08)	505,512	93,665	24,004	98,799	13,670	3,480	25,540	1,957	289
2008	70 (08-10)	640,947	137,103	39,420	125,311	18,738	5,200	33,419	2,797	552
2008	75 (06-08)	609,392	126,165	35,572	119,100	17,478	4,819	31,541	2,613	460
2008	75 (08-10)	730,567	169,117	53,037	140,977	22,664	6,605	38,606	3,440	683
2008	base	750,014	178,374	57,383	144,811	23,859	6,789	39,774	3,637	775
2009	60 (08-10)	207,026	23,833	3,939	40,838	4,491	1,011	9,625	368	13
2009	65 (08-10)	336,631	53,181	11,766	64,605	7,958	1,851	16,230	959	39
2009	70 (08-10)	404,703	72,497	18,016	77,290	10,019	2,377	19,618	1,313	79
2009	75 (08-10)	446,907	88,347	23,649	82,779	11,529	2,823	21,877	1,602	105
2009	base	455,048	92,536	25,593	83,908	11,936	3,046	22,192	1,641	118
2010	60 (08-10)	240,590	28,547	4,648	45,841	4,990	1,142	10,492	525	26
2010	65 (08-10)	409,772	66,877	15,784	74,467	9,428	2,390	18,265	1,234	92
2010	70 (08-10)	506,969	95,214	25,369	91,721	12,921	3,191	23,111	1,773	236
2010	75 (08-10)	579,467	118,667	34,167	103,014	15,613	3,952	26,525	2,272	328
2010	base	596,393	125,364	37,595	105,522	16,572	4,268	27,549	2,337	368

Counts of person-days with lung function decrement > 10,15,20%, study area=Detroit

year	scenario	FEV10 ages 5to18	FEV15 ages 5to18	FEV20 ages 5to18	FEV10 ages 19to35	FEV15 ages 19to35	FEV20 ages 19to35	FEV10 ages 36to55	FEV15 ages 36to55	FEV20 ages 36to55
2006	60 (06-08)	219,845	24,028	4,321	40,238	4,664	846	10,311	480	23
2006	65 (06-08)	301,189	42,067	8,711	54,207	7,133	1,738	14,563	1,075	160
2006	70 (06-08)	376,155	60,494	14,426	67,696	9,465	2,401	18,267	1,532	274
2006	75 (06-08)	477,641	90,124	24,920	87,037	13,877	3,772	24,531	2,241	412
2006	base	622,017	142,158	46,959	109,739	20,873	6,150	33,219	3,704	709
2007	60 (06-08)	284,042	35,277	6,767	41,655	4,550	983	10,745	594	114
2007	65 (06-08)	418,816	66,895	16,027	61,317	7,430	1,875	16,918	1,143	251
2007	70 (06-08)	544,354	100,526	27,709	81,070	10,974	3,064	23,205	1,829	412
2007	75 (06-08)	729,013	158,802	50,892	110,082	17,650	4,572	33,699	3,041	617
2007	base	1,039,530	283,585	105,167	163,238	31,459	9,419	54,641	5,738	1,257
2008	60 (06-08)	204,733	19,913	3,086	35,963	3,269	640	9,602	457	23
2008	60 (08-10)	264,655	31,024	5,807	44,902	4,641	846	11,911	754	46
2008	65 (06-08)	283,768	34,957	6,767	48,194	5,098	1,075	12,826	869	46
2008	65 (08-10)	385,483	58,848	13,695	63,923	7,819	1,783	17,124	1,235	137
2008	70 (06-08)	353,750	51,509	11,363	59,054	6,813	1,600	15,684	1,052	91
2008	70 (08-10)	506,059	91,884	24,554	83,836	11,728	2,835	23,525	1,966	251
2008	75 (06-08)	456,768	77,801	19,662	75,286	10,174	2,446	20,530	1,669	183
2008	75 (08-10)	575,789	120,439	36,854	91,427	14,975	3,864	26,269	2,561	343
2008	base	575,789	120,439	36,854	91,427	14,975	3,864	26,269	2,561	343
2009	60 (08-10)	210,334	24,234	4,435	40,055	4,435	846	10,357	526	23
2009	65 (08-10)	298,766	43,347	9,717	54,733	7,087	1,463	14,906	914	137
2009	70 (08-10)	384,020	64,746	17,238	69,502	10,037	2,309	19,319	1,463	229
2009	75 (08-10)	416,599	82,419	24,531	72,039	11,660	2,858	20,142	1,829	320
2009	base	416,599	82,419	24,531	72,039	11,660	2,858	20,142	1,829	320
2010	60 (08-10)	306,905	36,328	6,699	52,218	6,424	1,738	13,169	594	69
2010	65 (08-10)	449,200	68,130	15,249	73,503	10,471	2,561	18,930	1,486	251
2010	70 (08-10)	588,661	106,813	27,343	95,336	14,975	4,298	26,429	2,332	320
2010	75 (08-10)	682,351	144,399	41,587	106,927	18,427	5,601	30,270	2,835	503
2010	base	682,351	144,399	41,587	106,927	18,427	5,601	30,270	2,835	503

Counts of person-days with lung function decrement > 10,15,20%, study area=Houston

year	scenario	FEV10 ages 5to18	FEV15 ages 5to18	FEV20 ages 5to18	FEV10 ages 19to35	FEV15 ages 19to35	FEV20 ages 19to35	FEV10 ages 36to55	FEV15 ages 36to55	FEV20 ages 36to55
2006	60 (06-08)	318,714	30,694	4,671	53,184	4,494	491	10,205	412	59
2006	65 (06-08)	449,772	60,269	12,815	70,514	7,202	1,158	14,385	667	137
2006	70 (06-08)	556,946	88,490	22,353	85,782	9,950	1,707	18,840	1,060	196
2006	75 (06-08)	675,796	122,756	34,815	103,896	13,522	2,826	24,512	1,590	236
2006	base	1,248,677	364,795	145,188	201,022	40,389	12,089	60,976	7,458	1,668
2007	60 (06-08)	253,617	20,587	3,081	45,629	4,023	707	8,851	177	39
2007	65 (06-08)	336,887	36,385	6,476	57,090	5,436	1,178	11,500	412	39
2007	70 (06-08)	401,984	51,575	11,402	66,137	7,026	1,590	13,581	687	39
2007	75 (06-08)	473,244	69,591	16,995	77,481	8,988	2,021	16,485	981	98
2007	base	801,673	198,627	68,060	126,681	22,785	6,614	32,794	3,081	628
2008	60 (06-08)	269,318	22,255	3,081	48,200	4,651	765	9,930	294	39
2008	60 (08-10)	343,187	33,932	5,554	58,169	6,339	1,119	12,639	412	59
2008	65 (06-08)	352,980	35,973	6,005	59,563	6,594	1,138	12,972	412	59
2008	65 (08-10)	469,868	63,684	13,600	77,638	9,459	1,884	17,879	824	79
2008	70 (06-08)	415,526	50,241	9,636	69,061	8,046	1,531	15,543	628	59
2008	70 (08-10)	583,028	93,868	23,354	95,300	12,560	3,062	22,785	1,276	157
2008	75 (06-08)	482,134	66,589	14,366	78,952	9,773	2,002	18,467	864	98
2008	75 (08-10)	697,855	130,076	37,151	113,532	16,387	4,376	28,182	1,923	275
2008	base	750,745	166,638	53,126	118,497	19,213	5,436	31,518	3,160	530
2009	60 (08-10)	332,903	35,659	6,339	57,934	6,202	1,236	11,775	530	39
2009	65 (08-10)	448,869	65,352	15,759	74,792	9,302	1,727	16,740	1,099	118
2009	70 (08-10)	551,058	96,125	27,593	90,394	12,305	2,963	21,647	1,629	216
2009	75 (08-10)	651,029	131,234	41,939	105,545	16,760	4,357	27,044	2,512	412
2009	base	680,290	166,599	61,486	106,840	19,979	6,280	30,262	3,513	844
2010	60 (08-10)	354,962	38,250	6,908	60,269	6,123	922	12,246	510	59
2010	65 (08-10)	485,294	68,041	15,681	77,461	8,871	1,629	16,603	981	118
2010	70 (08-10)	601,004	101,247	26,828	93,907	11,736	2,551	21,490	1,472	236
2010	75 (08-10)	714,792	139,457	40,664	110,392	15,484	3,690	26,887	2,002	353
2010	base	772,824	179,728	60,446	116,162	19,429	5,004	30,851	2,846	530

Counts of person-days with lung function decrement > 10,15,20%, study area=LosAngeles

year	scenario	FEV10 ages 5to18	FEV15 ages 5to18	FEV20 ages 5to18	FEV10 ages 19to35	FEV15 ages 19to35	FEV20 ages 19to35	FEV10 ages 36to55	FEV15 ages 36to55	FEV20 ages 36to55
2006	60 (06-08)	1,836,948	216,331	39,245	446,940	40,814	9,269	92,169	4,186	374
2006	65 (06-08)	2,434,140	342,961	76,994	610,796	61,969	14,950	127,302	6,728	673
2006	70 (06-08)	3,065,119	496,800	126,106	777,044	91,421	21,678	165,874	10,764	1,570
2006	75 (06-08)	3,816,299	700,349	196,896	982,312	129,769	30,648	220,817	17,043	2,766
2006	base	5,668,272	1,699,330	691,005	1,415,573	291,905	100,317	421,375	61,072	17,342
2007	60 (06-08)	1,734,165	199,737	34,311	439,316	42,534	8,297	83,498	3,887	224
2007	65 (06-08)	2,297,718	317,246	66,828	585,231	64,660	13,605	117,734	6,055	523
2007	70 (06-08)	2,877,641	454,565	110,109	741,163	87,385	21,005	157,352	8,970	1,047
2007	75 (06-08)	3,588,680	634,791	174,695	944,338	122,443	31,097	208,482	13,754	1,944
2007	base	4,694,257	1,307,856	516,534	1,142,430	217,976	68,846	322,255	38,647	9,344
2008	60 (06-08)	1,945,712	230,908	42,459	474,449	44,029	8,746	94,262	5,158	374
2008	60 (08-10)	2,002,000	243,093	45,225	489,773	46,271	9,269	97,850	5,233	449
2008	65 (06-08)	2,582,895	360,154	80,582	637,782	68,547	14,726	132,460	7,998	822
2008	65 (08-10)	2,651,592	377,720	84,469	656,544	71,762	15,548	137,468	8,297	897
2008	70 (06-08)	3,247,139	517,282	129,320	813,373	98,597	21,977	177,685	11,661	1,420
2008	70 (08-10)	3,354,931	547,257	137,767	843,947	104,129	22,874	184,861	12,633	1,570
2008	75 (06-08)	4,046,235	730,100	201,755	1,037,628	141,430	33,339	234,421	17,193	2,766
2008	75 (08-10)	4,080,546	740,341	204,820	1,046,300	142,776	33,937	236,813	17,567	2,766
2008	base	5,829,736	1,683,035	677,400	1,488,680	294,970	92,842	433,784	56,139	12,783
2009	60 (08-10)	1,991,909	236,888	42,085	498,444	47,766	9,344	102,335	4,784	523
2009	65 (08-10)	2,615,487	375,478	80,582	660,955	73,406	17,193	139,038	8,447	822
2009	70 (08-10)	3,300,661	541,576	132,684	849,105	105,848	25,565	185,309	11,736	1,346
2009	75 (08-10)	4,000,860	725,615	199,064	1,057,064	141,281	35,582	237,112	16,670	2,915
2009	base	5,227,013	1,415,573	539,483	1,269,957	236,963	73,855	358,285	47,542	10,390
2010	60 (08-10)	1,602,078	177,909	29,153	440,886	42,235	8,970	84,768	4,859	822
2010	65 (08-10)	2,112,408	277,628	56,512	579,027	63,614	14,577	119,154	7,400	972
2010	70 (08-10)	2,657,124	398,053	96,878	746,545	89,029	20,856	162,510	10,390	1,719
2010	75 (08-10)	3,219,705	543,520	142,103	918,250	119,005	30,573	205,418	14,203	2,841
2010	base	3,494,119	848,656	300,502	844,171	132,086	36,404	232,328	26,911	3,663

Counts of person-days with lung function decrement > 10,15,20%, study area=NewYork

year	scenario	FEV10 ages 5to18	FEV15 ages 5to18	FEV20 ages 5to18	FEV10 ages 19to35	FEV15 ages 19to35	FEV20 ages 19to35	FEV10 ages 36to55	FEV15 ages 36to55	FEV20 ages 36to55
2006	65 (06-08)	192,061	9,631	1,111	42,783	3,519	370	9,075	278	0
2006	70 (06-08)	1,114,678	156,594	35,838	180,115	18,706	3,982	41,857	1,945	278
2006	75 (06-08)	1,673,916	296,334	84,826	265,497	32,504	7,871	68,064	4,075	648
2006	base	3,473,774	995,774	384,771	601,743	118,534	39,172	187,894	19,447	4,075
2007	65 (06-08)	232,344	16,391	2,593	52,229	4,352	556	11,576	741	93
2007	70 (06-08)	1,355,357	199,840	46,950	225,769	23,522	4,815	52,599	3,612	926
2007	75 (06-08)	1,999,883	375,048	98,253	325,782	40,005	9,168	77,325	5,186	1,111
2007	base	3,914,663	1,104,307	415,608	670,733	135,017	39,172	195,951	18,243	3,797
2008	65 (06-08)	208,638	12,224	2,315	44,821	3,612	648	8,890	463	0
2008	65 (08-10)	355,138	24,911	4,445	72,231	6,482	1,111	15,187	741	93
2008	70 (06-08)	1,168,204	170,670	37,783	209,656	23,799	5,464	51,488	3,149	556
2008	70 (08-10)	1,543,807	259,292	71,213	269,571	34,264	8,705	68,342	3,797	833
2008	75 (06-08)	1,716,607	300,686	87,882	297,260	39,449	10,464	76,028	4,445	1,019
2008	75 (08-10)	2,392,896	498,860	167,521	412,182	62,786	18,613	112,885	9,723	1,574
2008	base	3,280,324	869,555	323,282	572,387	109,458	36,394	172,614	17,780	3,334
2009	65 (08-10)	274,387	21,392	2,500	63,341	4,630	833	14,076	463	0
2009	70 (08-10)	943,360	133,721	29,541	171,966	20,373	3,612	43,709	2,593	370
2009	75 (08-10)	1,372,582	237,715	63,249	241,883	32,597	8,149	64,453	5,093	648
2009	base	1,671,138	354,860	111,866	274,572	41,857	12,039	77,880	7,408	833
2010	65 (08-10)	422,646	33,245	5,186	87,789	7,316	926	18,984	648	0
2010	70 (08-10)	2,033,313	345,785	95,197	328,745	39,079	9,075	85,752	5,278	556
2010	75 (08-10)	3,184,571	686,939	214,286	509,694	77,510	19,262	144,555	9,816	2,408
2010	base	4,291,285	1,142,182	419,127	689,254	130,202	36,579	207,804	19,169	4,075

Counts of person-days with lung function decrement > 10,15,20%, study area=Philadelphia

year	scenario	FEV10 ages 5to18	FEV15 ages 5to18	FEV20 ages 5to18	FEV10 ages 19to35	FEV15 ages 19to35	FEV20 ages 19to35	FEV10 ages 36to55	FEV15 ages 36to55	FEV20 ages 36to55
2006	60 (06-08)	372,132	42,431	7,242	46,727	4,130	771	13,162	771	110
2006	65 (06-08)	500,610	68,038	14,043	62,146	6,305	1,156	18,173	1,074	193
2006	70 (06-08)	665,626	109,396	26,323	82,467	9,417	2,010	25,580	1,542	275
2006	75 (06-08)	842,839	159,867	44,248	106,973	13,547	3,166	33,014	2,313	413
2006	base	1,542,222	424,614	156,921	214,496	38,576	10,959	76,684	7,159	1,459
2007	60 (06-08)	445,898	57,107	11,399	54,078	4,901	771	14,786	909	110
2007	65 (06-08)	596,348	91,388	22,799	74,729	7,985	1,597	21,450	1,377	248
2007	70 (06-08)	797,655	144,145	40,118	101,548	12,831	2,671	30,040	2,203	441
2007	75 (06-08)	1,007,139	208,273	62,724	131,010	18,063	4,571	40,862	3,194	688
2007	base	1,795,900	513,303	197,149	255,192	51,049	15,970	94,940	10,298	2,671
2008	60 (06-08)	317,008	32,381	4,708	40,889	3,910	606	11,785	523	193
2008	60 (08-10)	425,522	53,115	10,436	53,996	5,672	854	15,888	909	220
2008	65 (06-08)	425,522	53,115	10,436	53,996	5,672	854	15,888	909	220
2008	65 (08-10)	572,310	87,561	20,513	73,380	8,288	2,010	21,422	1,652	330
2008	70 (06-08)	572,310	87,561	20,513	73,380	8,288	2,010	21,422	1,652	330
2008	70 (08-10)	768,027	143,676	39,320	101,108	13,905	3,524	31,307	2,698	523
2008	75 (06-08)	723,421	130,377	34,556	94,472	12,556	3,084	28,884	2,423	496
2008	75 (08-10)	994,831	213,202	66,111	133,791	20,706	5,507	43,588	3,993	909
2008	base	1,395,710	367,617	136,765	198,718	36,566	10,849	69,415	7,958	1,625
2009	60 (08-10)	267,445	26,957	4,598	36,676	3,359	496	10,133	468	28
2009	65 (08-10)	335,621	41,275	7,930	45,487	4,598	743	12,804	743	55
2009	70 (08-10)	429,295	62,229	13,602	58,264	6,250	1,212	16,356	1,019	138
2009	75 (08-10)	530,870	87,891	21,670	72,527	8,839	1,955	20,486	1,514	193
2009	base	696,437	141,033	42,872	92,820	13,162	3,359	27,287	2,340	358
2010	60 (08-10)	514,350	66,331	13,630	60,494	5,755	606	17,485	1,019	110
2010	65 (08-10)	683,413	107,551	25,800	81,283	8,894	1,707	24,451	1,762	138
2010	70 (08-10)	916,825	171,074	46,782	113,085	14,841	3,166	35,272	2,753	358
2010	75 (08-10)	1,168,163	247,565	77,263	150,037	21,945	5,177	48,654	3,937	799
2010	base	1,600,266	414,453	148,550	215,184	37,723	11,262	74,757	7,159	1,569

Counts of person-days with lung function decrement > 10,15,20%, study area=Sacramento

year	scenario	FEV10 ages 5to18	FEV15 ages 5to18	FEV20 ages 5to18	FEV10 ages 19to35	FEV15 ages 19to35	FEV20 ages 19to35	FEV10 ages 36to55	FEV15 ages 36to55	FEV20 ages 36to55
2006	60 (06-08)	170,427	21,819	4,807	19,160	1,628	144	4,479	202	10
2006	65 (06-08)	252,038	41,065	10,433	29,506	2,977	539	7,379	453	48
2006	70 (06-08)	319,680	58,761	16,521	38,233	4,325	819	10,144	578	87
2006	75 (06-08)	402,611	83,162	25,508	49,542	6,473	1,378	13,939	1,031	173
2006	base	847,289	254,234	100,087	118,707	23,370	7,273	40,170	5,289	1,060
2007	60 (06-08)	113,255	11,752	2,495	13,987	934	135	3,034	135	0
2007	65 (06-08)	161,141	21,193	5,183	19,256	1,609	222	4,373	212	19
2007	70 (06-08)	199,046	29,785	7,793	23,639	2,158	414	5,606	289	19
2007	75 (06-08)	245,950	41,537	11,656	29,564	3,217	636	7,273	443	39
2007	base	480,946	118,283	42,501	61,410	10,384	2,678	18,813	1,772	328
2008	60 (06-08)	162,065	20,827	4,720	19,401	1,580	337	4,422	125	0
2008	60 (08-10)	162,065	20,827	4,720	19,401	1,580	337	4,422	125	0
2008	65 (06-08)	236,143	38,243	10,038	28,571	3,005	617	6,897	347	19
2008	65 (08-10)	226,674	35,748	9,209	27,223	2,774	549	6,522	308	10
2008	70 (06-08)	296,667	54,388	15,182	36,702	4,316	867	9,325	617	29
2008	70 (08-10)	286,090	51,710	14,209	35,305	4,075	800	8,737	549	29
2008	75 (06-08)	372,566	76,139	23,803	47,240	6,367	1,339	12,937	973	106
2008	75 (08-10)	361,584	72,787	22,358	45,603	6,136	1,272	12,292	934	87
2008	base	766,690	222,609	89,298	111,868	20,904	6,599	36,201	4,701	1,060
2009	60 (08-10)	142,655	17,359	3,400	17,118	1,474	222	3,795	231	29
2009	65 (08-10)	198,767	29,959	7,562	23,793	2,524	511	5,578	328	39
2009	70 (08-10)	249,466	42,867	11,955	30,431	3,555	751	7,629	511	77
2009	75 (08-10)	315,153	60,929	18,206	39,062	5,269	1,175	10,307	771	116
2009	base	649,408	181,640	70,687	89,905	17,571	5,606	28,292	3,179	761
2010	60 (08-10)	98,372	9,296	1,753	12,667	828	87	2,803	106	0
2010	65 (08-10)	133,648	16,328	3,449	16,790	1,214	164	3,872	125	10
2010	70 (08-10)	168,086	22,724	5,626	20,412	1,734	222	5,038	241	29
2010	75 (08-10)	209,951	32,271	9,007	25,470	2,428	385	6,550	347	48
2010	base	417,783	100,588	34,997	52,683	7,774	1,859	16,126	1,512	318

Counts of person-days with lung function decrement > 10,15,20%, study area=StLouis

year	scenario	FEV10 ages 5to18	FEV15 ages 5to18	FEV20 ages 5to18	FEV10 ages 19to35	FEV15 ages 19to35	FEV20 ages 19to35	FEV10 ages 36to55	FEV15 ages 36to55	FEV20 ages 36to55
2006	60 (06-08)	150,775	17,506	3,230	21,566	2,200	456	5,453	293	47
2006	65 (06-08)	228,252	34,684	7,805	33,127	3,943	901	8,811	620	82
2006	70 (06-08)	327,493	60,415	15,973	48,620	6,635	1,744	13,539	1,135	222
2006	75 (06-08)	440,753	93,835	28,283	66,664	10,754	2,879	19,191	1,767	386
2006	base	655,057	171,359	60,930	100,236	19,319	5,593	33,361	3,745	854
2007	60 (06-08)	183,189	25,685	5,313	24,176	2,340	538	6,483	410	35
2007	65 (06-08)	279,435	48,609	12,685	38,335	4,400	1,018	10,251	807	164
2007	70 (06-08)	397,715	80,847	24,761	56,062	7,957	1,884	15,996	1,369	351
2007	75 (06-08)	531,476	124,575	42,278	77,839	12,860	3,440	23,953	2,375	608
2007	base	778,767	219,148	82,344	118,058	23,918	7,126	40,417	4,727	1,252
2008	60 (06-08)	83,128	7,501	924	13,433	1,240	222	3,031	105	23
2008	60 (08-10)	134,967	15,844	2,808	20,490	2,235	550	5,313	211	23
2008	65 (06-08)	121,170	13,410	2,153	18,500	1,942	433	4,669	176	23
2008	65 (08-10)	196,833	28,318	6,553	29,956	3,464	831	8,027	410	59
2008	70 (06-08)	168,480	22,163	4,751	25,708	2,867	679	6,623	304	47
2008	70 (08-10)	256,617	43,249	11,315	38,206	5,079	1,275	11,070	726	140
2008	75 (06-08)	219,394	33,829	8,226	33,385	4,002	960	9,069	503	82
2008	75 (08-10)	298,274	55,208	15,107	43,659	6,085	1,556	13,012	1,030	164
2008	base	310,046	59,046	16,698	45,367	6,448	1,627	13,609	1,077	152
2009	60 (08-10)	109,668	12,708	2,223	16,160	1,533	316	4,084	187	12
2009	65 (08-10)	159,891	22,760	4,997	23,345	2,516	550	6,026	386	47
2009	70 (08-10)	208,464	35,222	9,174	30,424	3,522	866	8,179	585	94
2009	75 (08-10)	249,338	46,385	12,942	35,526	4,622	1,229	9,970	866	117
2009	base	260,759	50,574	14,674	36,989	4,961	1,404	10,742	924	129
2010	60 (08-10)	179,643	24,339	5,032	26,399	3,206	702	7,583	398	23
2010	65 (08-10)	268,424	45,952	11,491	39,856	5,535	1,404	11,936	878	94
2010	70 (08-10)	357,180	69,765	19,904	53,593	8,671	2,282	16,499	1,556	234
2010	75 (08-10)	428,759	93,239	29,067	63,938	11,093	3,276	20,513	2,083	433
2010	base	450,454	101,242	32,577	67,308	11,936	3,628	22,034	2,200	468

Counts of person-days with lung function decrement > 10,15,20%, study area=WashingtonDC

year	scenario	FEV10 ages 5to18	FEV15 ages 5to18	FEV20 ages 5to18	FEV10 ages 19to35	FEV15 ages 19to35	FEV20 ages 19to35	FEV10 ages 36to55	FEV15 ages 36to55	FEV20 ages 36to55
2006	60 (06-08)	247,748	29,307	5,600	43,747	5,106	1,125	10,189	585	90
2006	65 (06-08)	364,391	54,813	12,640	63,292	7,895	1,822	14,710	1,147	135
2006	70 (06-08)	497,498	88,955	23,437	88,101	11,246	2,901	21,435	1,822	292
2006	75 (06-08)	647,631	131,510	39,181	114,394	16,307	4,453	28,655	2,564	495
2006	base	1,298,433	372,443	144,960	234,658	47,188	14,170	70,827	8,547	1,867
2007	60 (06-08)	306,564	36,999	6,500	53,936	5,488	1,035	12,505	765	90
2007	65 (06-08)	444,079	67,476	14,237	77,147	9,402	1,664	18,961	1,439	180
2007	70 (06-08)	601,500	106,994	27,418	103,980	14,327	3,104	26,203	2,204	360
2007	75 (06-08)	769,492	154,587	44,129	133,129	20,355	5,151	34,705	3,171	675
2007	base	1,479,875	409,914	151,055	257,509	50,922	15,902	76,562	9,109	2,317
2008	60 (06-08)	209,759	21,817	3,576	38,281	4,116	832	8,322	315	22
2008	60 (08-10)	244,936	27,575	5,061	43,612	4,858	1,012	9,447	450	22
2008	65 (06-08)	296,780	38,236	7,647	52,249	6,073	1,305	11,718	607	67
2008	65 (08-10)	395,295	60,728	14,372	68,488	8,434	2,204	16,352	990	180
2008	70 (06-08)	395,295	60,728	14,372	68,488	8,434	2,204	16,352	990	180
2008	70 (08-10)	547,924	102,136	28,407	93,161	13,180	3,891	24,696	1,619	382
2008	75 (06-08)	503,008	89,360	23,976	85,582	11,898	3,126	22,177	1,417	292
2008	75 (08-10)	768,795	169,836	55,105	131,218	21,120	6,095	37,359	3,014	697
2008	base	957,591	239,246	84,997	160,277	28,362	8,569	48,965	4,341	922
2009	60 (08-10)	168,936	15,924	2,722	32,073	3,284	540	7,332	405	90
2009	65 (08-10)	250,807	30,342	6,253	44,331	4,903	1,102	11,066	607	135
2009	70 (08-10)	328,584	47,390	11,268	56,769	6,703	1,484	14,732	1,035	202
2009	75 (08-10)	428,920	73,683	19,276	71,974	9,064	2,294	19,276	1,574	270
2009	base	500,242	97,682	27,260	78,856	11,133	2,654	22,402	1,912	382
2010	60 (08-10)	345,048	44,286	8,479	59,648	6,703	1,462	14,845	900	157
2010	65 (08-10)	559,890	94,983	23,594	97,232	13,158	3,036	25,978	2,002	360
2010	70 (08-10)	774,890	155,554	45,299	136,458	20,558	5,758	37,292	3,554	765
2010	75 (08-10)	1,079,295	252,134	84,030	189,921	32,816	9,604	55,083	5,645	1,237
2010	base	1,325,221	347,724	124,785	230,362	44,286	13,810	69,522	7,917	1,777

1
2
3

1

2

3 **Appendix 6C**

4 **Comparison of E-R Function Risk Results With the 2007 Review**

5

1   In this appendix we compare lung function risk estimates for school-aged children for

2   this review and the previous review, based on the population exposure-response (E-R) function

3   methodology used in previous reviews.

4   Of the alternative standards for which risk results are available for the 2007 O_3 NAAQS

5   review, the standard with level 0.074 ppm with the same form as the existing standard (which

6   has level 0.075) is the most comparable for comparison with risks estimated to remain after just

7   meeting the existing standard of 0.075 ppm. Table 6C-1 compares the estimated percents of

8   asthmatic school-aged children with responses $\geq 10\%$ and the estimated percents of all school-

9   aged children with responses $\geq 15\%$ for these two scenarios. The results for Atlanta overlap, but

10   the existing standard results tend to be lower than the results for the 2007 review. The existing

11   standard results for Los Angeles are uniformly higher than the corresponding results for the 2007

12   review. In the previous review the response was extrapolated down to, and set to zero below,

13   background levels, which ranged from 0.013 to 0.033 ppm. For the existing standard analysis,

14   the response is calculated down to exposure concentrations of zero. We do not expect this to

15   result in much difference; the exposure response function is close to zero at levels at or near

16   background levels (see Table 6A-1 in Appendix 6A), However, while the level of the standard

17   evaluated for each scenario is almost the same, the starting distribution of air quality from which

18   attainment was simulated is different, since the last review used different years of air quality

19   data. This will result in some differences.

20

1

Table 6C-1. Comparison of existing standard results with the 2007 O3 NAAQS review,
using the population exposure-response method: Ranges (over years) of responses[a] for
asthmatic and all school-aged children

	FEV$_1$ decrement ≥ 10%, asthmatic school-aged children				FEV$_1$ decrement ≥ 15%, all school-aged children			
	Existing 0.075 ppm standard, 2006-2010		2007 review 0.074 ppm standard, 2002-2004		Existing 0.075 ppm standard, 2006-2010		2007 review 0.074 ppm standard, 2002-2004	
Urban area	min	max	min	max	min	max	min	max
Atlanta	3.8%	5.6%	4.6%	7.3%	0.9%	1.7%	1.2%	2.2%
Chicago	3.5%	6.3%	2.1%	6.5%	0.8%	2.0%	0.5%	2.0%
Houston	3.4%	5.3%	3.9%	4.4%	0.8%	1.6%	1.0%	1.2%
Los Angeles	4.3%	4.8%	1.9%	2.0%	1.0%	1.2%	0.5%	0.5%
New York	3.7%	5.7%	2.7%	6.6%	0.9%	1.7%	0.6%	2.0%

5 [a] Percents of school-aged children estimated to experience lung function responses associated
6 with 8-hour O$_3$ exposure while engaged in moderate or greater exertion. Results from the 2007
7 review are from Tables 3-23 and 3-27 in the 2007 Risk Assessment TSD.

8

1
2

1
2
3
4

Appendix 6D

5

6

Comparison of the MSS Model With Results From a Clinical Study With Children Ages 8-11

7

8

9

10

1   This appendix compares the predictions of the MSS model for 8-11 year old children

2   using the age term extension described in Section 6.2.4 with FEV_1 responses for this age group

3   reported by McDonnell et al. (1985).

4   Table 6D-1 gives the FEV_1 responses for a clinical study with children ages 8-11,

5   exposed to 120 ppb ozone over 2.5 hours at heavy exertion levels done by McDonnell et al.

6   (1985).

7

8   **Table 6D-1. FEV_1 responses measured during clean air and exposures to 120 ppb ozone[a]**

Subject	Clean air pre-exposure	Clean air post-exposure	Ozone pre-exposure	Ozone post-exposure	$-(CA\ post - CA\ pre)/CA\ pre$	$-(O_3\ post - O_3\ pre)/O_3\ pre$	CA-adjusted FEV1 decrement
1	2,256	2,274	2,324	2,299	−0.80%	1.08%	1.87%
2	1,669	1,832	1,831	1,780	−9.77%	2.79%	12.55%
3	1,954	1,935	1,949	1,845	0.97%	5.34%	4.36%
4	1,675	1,656	1,709	1,697	1.13%	0.70%	−0.43%
6	2,331	2,264	2,245	2,090	2.87%	6.90%	4.03%
7	2,511	2,490	2,409	2,541	0.84%	−5.48%	−6.32%
8	2,281	2,345	2,252	2,287	−2.81%	−1.55%	1.25%
9	1,937	1,911	2,061	1,919	1.34%	6.89%	5.55%
10	1,841	1,807	1,795	1,714	1.85%	4.51%	2.67%
11	1,431	1,686	1,615	1,513	−17.82%	6.32%	24.14%
12	2,251	2,122	2,179	2,147	5.73%	1.47%	−4.26%
13	1,824	1,789	1,803	1,700	1.92%	5.71%	3.79%
14	1,670	1,638	1,794	1,788	1.92%	0.33%	−1.58%
16	1,519	1,495	1,532	1,312	1.58%	14.36%	12.78%
17	1,542	1,465	1,524	1,191	4.99%	21.85%	16.86%
18	2,372	2,205	2,233	2,121	7.04%	5.02%	−2.02%
20	1,643	1,654	1,673	1,583	−0.67%	5.38%	6.05%
21	2,349	2,416	2,353	2,287	−2.85%	2.80%	5.66%
22	2,519	2,549	2,545	2,590	−1.19%	−1.77%	−0.58%
24	2,340	2,365	2,487	2,424	−1.07%	2.53%	3.60%
25	1,606	1,416	1,502	1,592	11.83%	−5.99%	−17.82%
26	1,824	1,801	1,836	1,759	1.26%	4.19%	2.93%

9   [a] from Table 2, McDonnell et al. (1985)

10

11   The numbers of subjects with clean-air adjusted responses greater than 10%, 15%, and 20% are

12   respectively 4, 2, and 1, corresponding to 18.2%, 9.1%, and 4.5% of the number of subjects.

13

1   Figure 6D-1 shows the distribution of clean-air corrected FEV_1 decrements across

2   subjects in the McDonnell et al. (1985) study (the last column in Table 6D-1). Figure 6D-2

3   displays the last 3 columns in Table 6D-1 and illustrates the variability of responses typical of

4   these studies of ozone exposure.

5

6   **Figure 6D-1. Clean-air Corrected FEV_1 Decrement vs. Subject id**

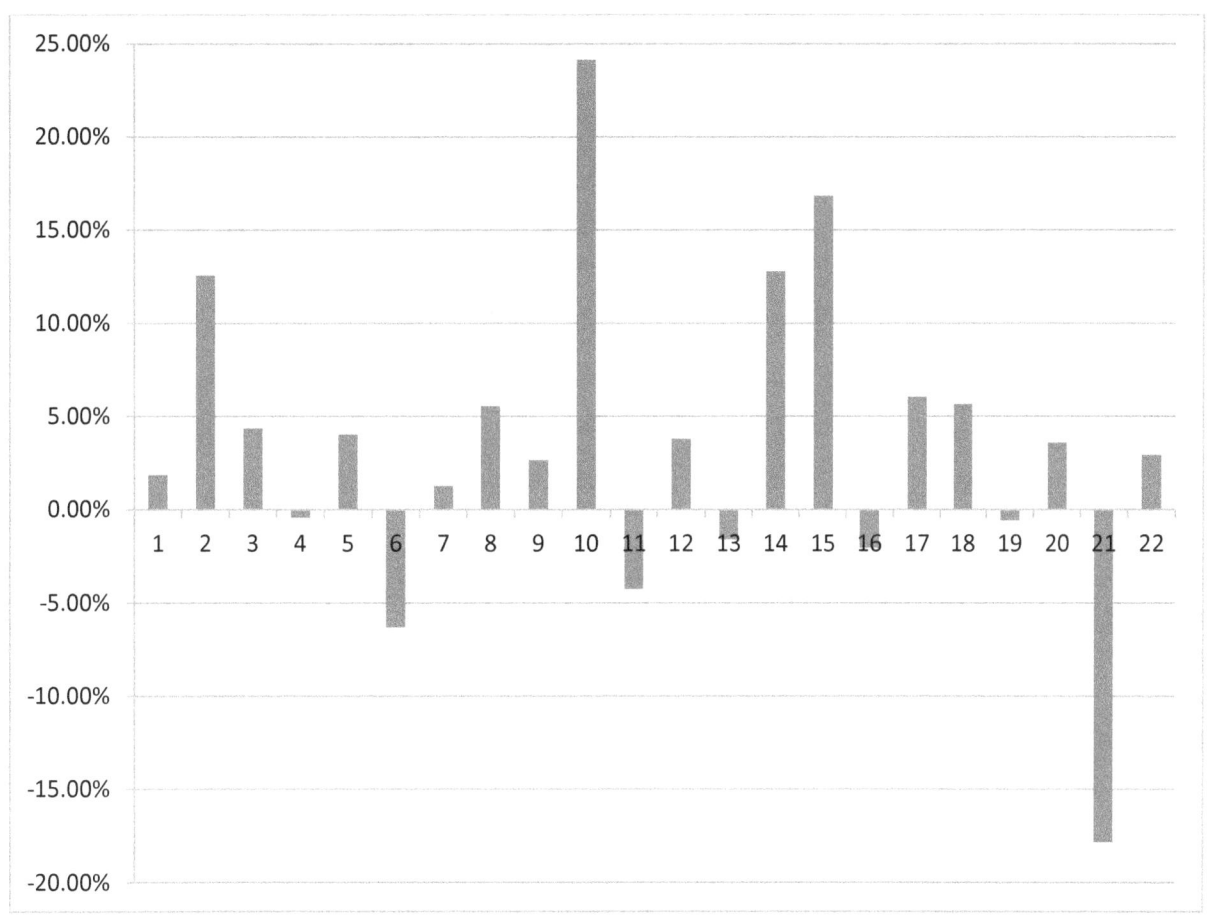

7

8

1 **Figure 6D-2. FEV₁ responses measured during clean air and exposures to 120 ppb ozone[a]**

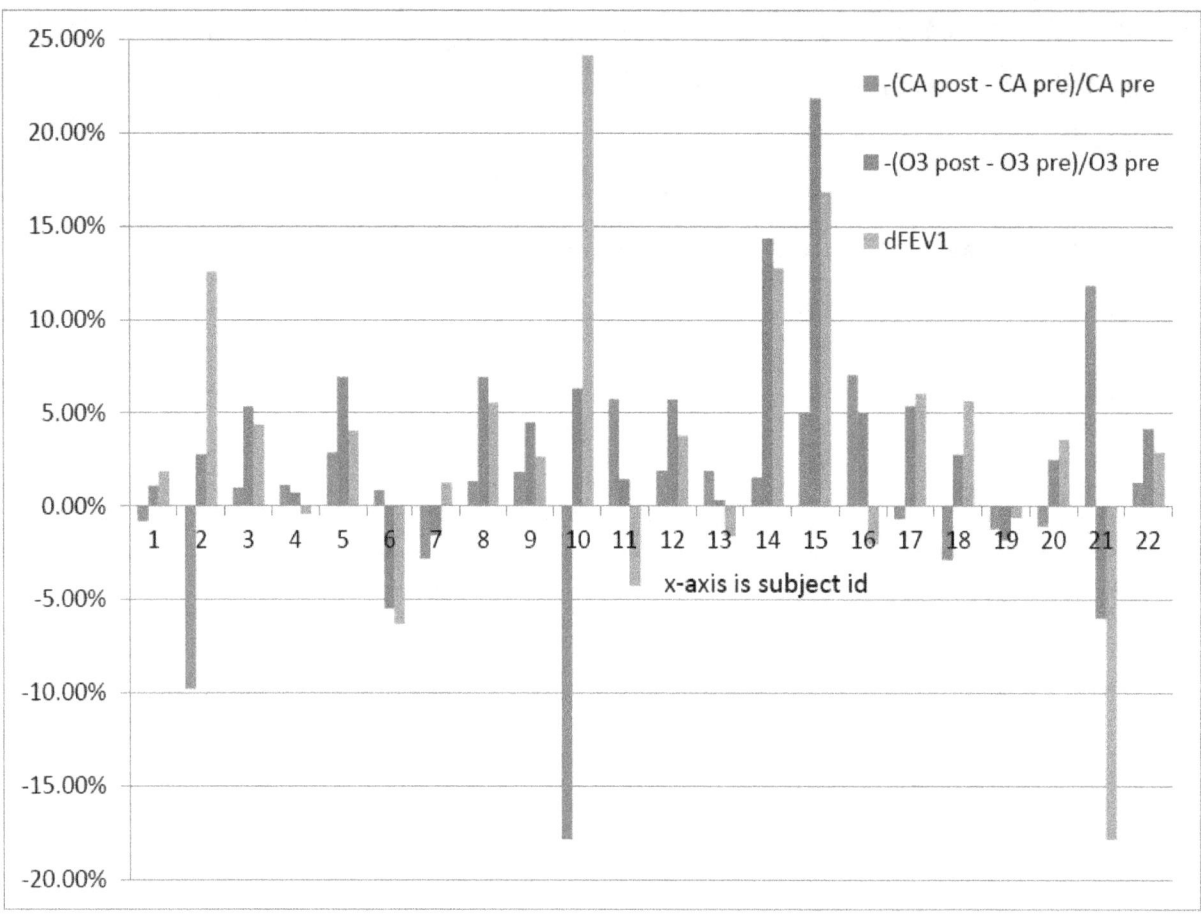

2

3 [a] derived from data in McDonnell et al. (1985)

4

5

6   Without ventilation rate measurements, this study cannot be used to fit the MSS model

7 for children. However, McDonnell et al. (1985) report mean values for ventilation rates

8 normalized by body surface area (BSA) of 32.4 (±3.3) for the clean air exposures and 33.3 (±3.4)

9 L/min/m² for the ozone exposures (standard deviations in parentheses). We can run the MSS

10 model with this study's exposure/exercise protocol for a sample of 8-11 year old children with

11 exercising ventilation rates sampled from Gaussian distributions with these means and standard

12 deviations, constraining the normalized ventilation rates to be within 1.5 standard deviations of

13 the mean. Resting ventilation rates were assumed to be 10.4 L/min (Avol et al., 1985) and BSA

1   to be 1.08 m^2 (EPA Exposure Factors Handbook[1]). Table 6D-2 compares the results of this

2   simulation with the results of McDonnell et al. (1985). The agreement is fairly good. Due to the

3   limited sample size of 22 subjects from only one study and the assumptions made in running the

4   MSS model, this does not provide confirmation that the age term extension is correct; however,

5   this comparison is supportive of the age term extension.

6

7

8   **Table 6D-2. Comparison of responses from the MSS model with responses from**

9   **McDonnell et al. (1985).**

	$\geq 10\%$ FEV$_1$ decrement		$\geq 15\%$ FEV$_1$ decrement		$\geq 20\%$ FEV$_1$ decrement	
	MSS model	**McDonnell et al. (1985)**	**MSS model**	**McDonnell et al. (1985)**	**MSS model**	**McDonnell et al. (1985)**
Percent responding	18.4%	18.2% (4 subjects)	6.8%	9.1% (2 subjects)	2.3%	4.5% (1 subject)

10

11

---

[1] U.S. EPA. Exposure Factors Handbook 2011 Edition (Final). U.S. Environmental Protection Agency, Washington, DC, EPA/600/R-09/052F, 2011.

1
2

1

2

3

Appendix 6E

4

5

Factors Related to Age

6

7

8

1   There are several factors related to the age of an individual that influence the risk of the

2   individual of experiencing FEV_1 decrements (ΔFEV_1) in excess of 10, 15, and 20%. These

3   factors include the parameters of the MSS model and the inputs to the MSS model calculated by

4   APEX: times series of exposures and ventilation rates normalized by body surface area (Figure

5   6E-1). Figure 6E-2 illustrates the combined effect of these factors on modeled FEV_1 decrements

6   across ages. Children are estimated to have the highest risk and older adults the lowest risk. In

7   this section we present sensitivity analyses which reveal the most influential underlying

8   components driving age-related differences in risk.

9

10   **Figure 6E-1. Factors That Contribute to Risk (FEV_1 Decrements)**

11

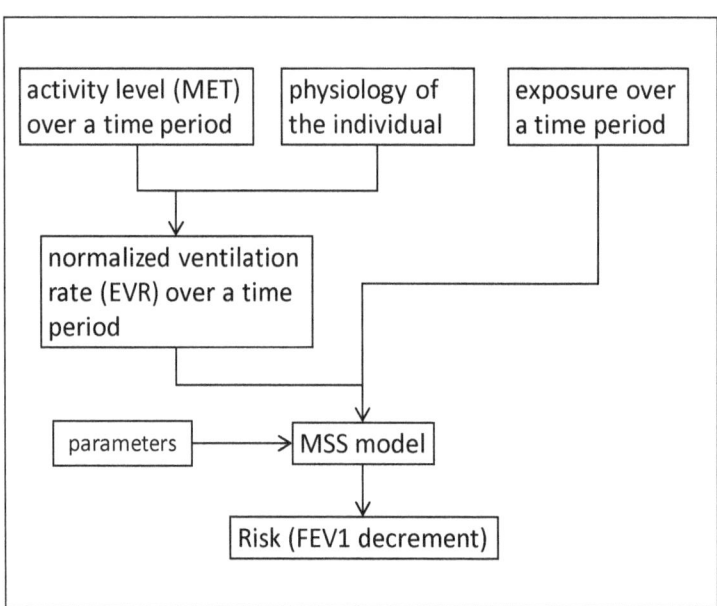

12

13

14

15   The times series of exposures modeled by APEX depend on ozone concentrations and

16   activity patterns. The times series of ventilation rates modeled by APEX depend on activity

17   patterns, exertion levels, and individual physiological characteristics. Figure 6E-3 provides an

18   overview of how APEX models these quantities to estimate risk (the acronyms are defined in

19   Table 6E-1). As this figure shows, there are several quantities input to APEX whose

20   distributions depend on age: MOXD, BM, VO2max, MET, BM, RMR, VE regression

21   coefficients, CHAD diary locations, and CHAD diary activities. These factors are

1   intercorrelated, so there is not a unique way to apportion their influences on FEV_1 decrements.

2   Since the highest exposures occur when an individual is outdoors, additional relevant factors

3   include the amount of time spent outdoors and the O_3 concentrations when individuals are

4   outdoors.

5

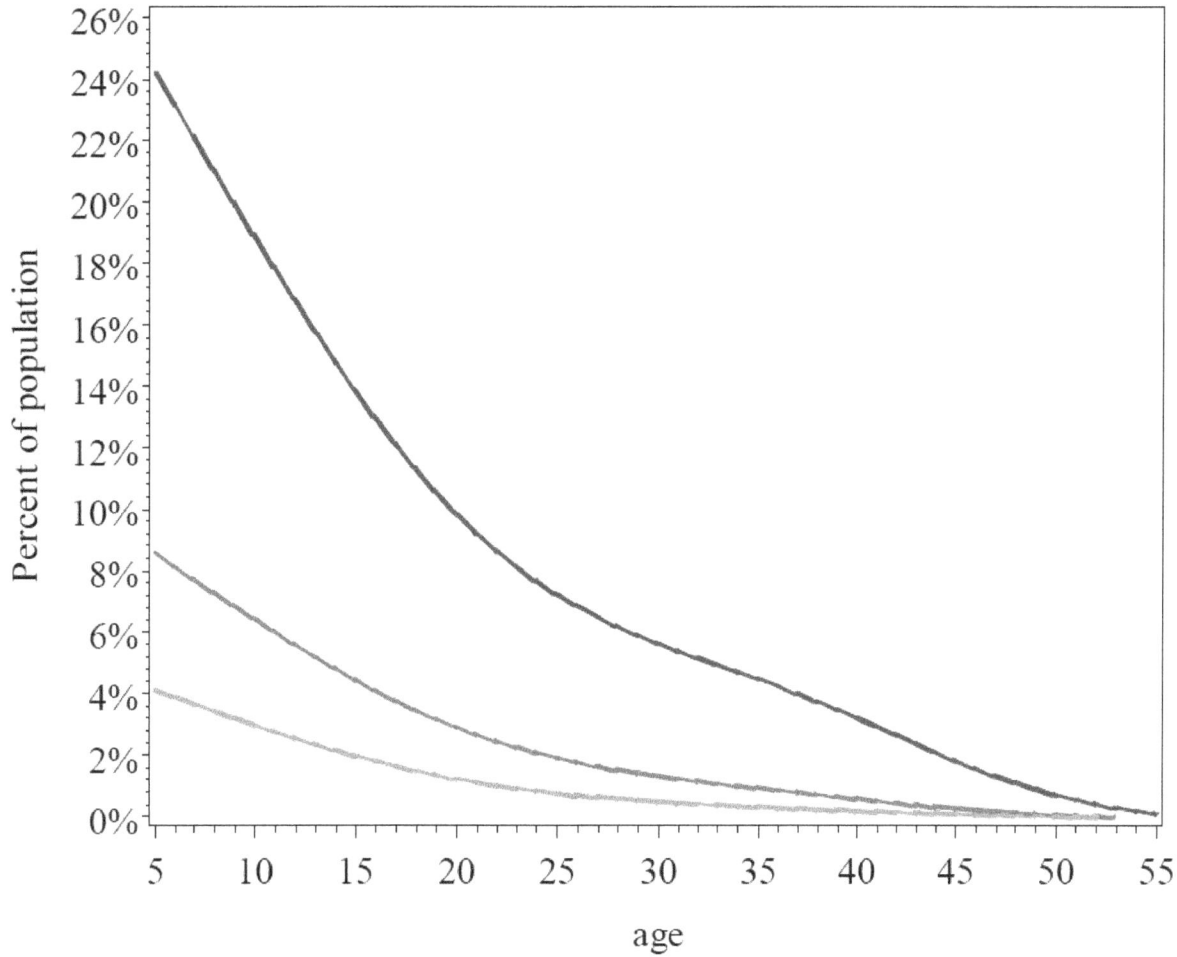

6

7   **Figure 6E-2. The Relationship Between Age and Percent of Population with Exceedances**
8   **of 10% (blue), 15% (green), and 20% (red) FEV_1 Decrements.** Based on the MSS (2012)
9   Threshold Model in an APEX Simulation of Atlanta for April-June 2006 (ages 5-95, 250,000
10   profiles).
11

Figure 6E-3. Overview Chart of APEX Modeling of Factors Input to the MSS Model

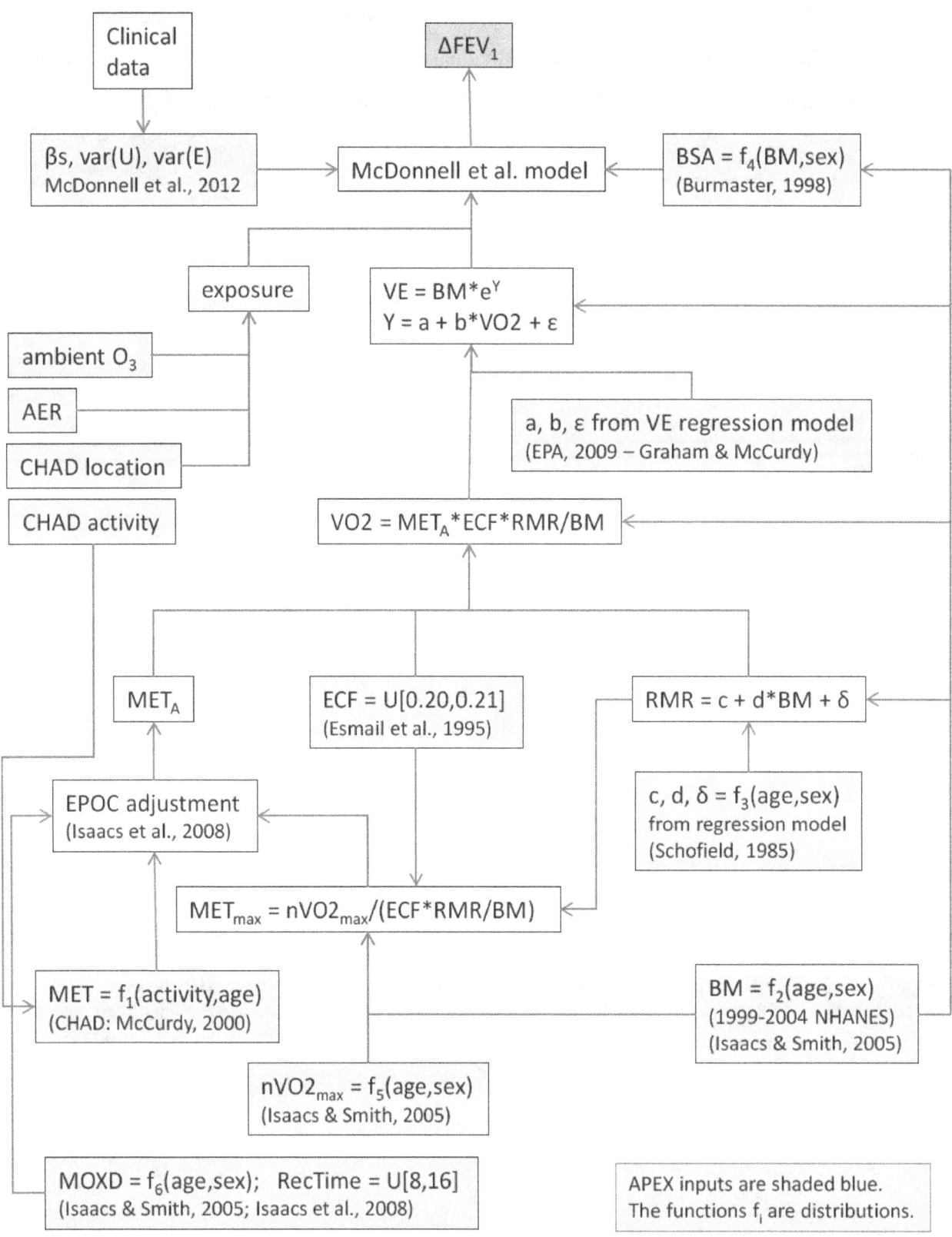

Table 6E-1. APEX Acronyms

Acronym	Description
AER	Air exchange rate (1/h)
BM	Body mass (kg)
BSA	Body Surface Area (m^2)
CHAD	Consolidated Human Activity Database
ECF	Energy conversion factor (L O_2/kcal)
EE	Energy expenditure (kcal/min)
EErest	Resting Energy expenditure (kcal/min)
EPOC	Excess post-exercise oxygen consumption
EVR	Equivalent ventilation rate. EVR = VE / BSA (L/min-m^2)
FEV_1	Forced expiratory volume in 1 second
MET	Metabolic equivalent of work. MET = EE / EErest (unitless)
MET_A	MET adjusted for EPOC
METmax	Maximum achievable MET for an individual (unitless)
MOXD	Maximal oxygen deficit that can be obtained (M-h)
RecTime	Oxygen debt recovery time (h). Time required to recover from F=1 to 0 while at rest, where F is the fractional oxygen deficit.
RMR	Resting metabolic rate (kcal/min)
VE	Ventilation rate (L/min)
VO_2	Oxygen consumption (L O_2/min)
VO_2max	Maximum oxygen consumption for an individual (L O_2/min)
nVO_2max	Maximum normalized oxygen consumption (L O_2/min/kg)

MSS Model Age Term

Figure 6E-4 illustrates the interaction of age and ozone level for the prediction of risk in the MSS threshold model. This figure assumes moderate exercising conditions of a typical 6.6-hour clinical study (EVR = 20 L/min-m^2 BSA) and varies the ozone exposure level (constant over the 6.6 hour period) from 40 to 120 ppb. The lung function decrement is the median value predicted at the end of the 6.6 hour period (U_i and ε_{ijk} = 0 in equation 6-3). The trend of the slope vs. age increasing with O$_3$ exposure concentrations is consistent with the findings of McDonnell et al. (1993, Figures 3 and 4). In our application of the model, the response is flat for ages below 18 and declines out to age 55, after which we assume no response (Figure 6E-5).

Figure 6E-4. Variation by Age of the Median Response (Lung Function Decrements in FEV$_1$) Predicted by the MSS Model After 6.6 Hour Exposure to Ozone Under Intermittent Moderate Exercise

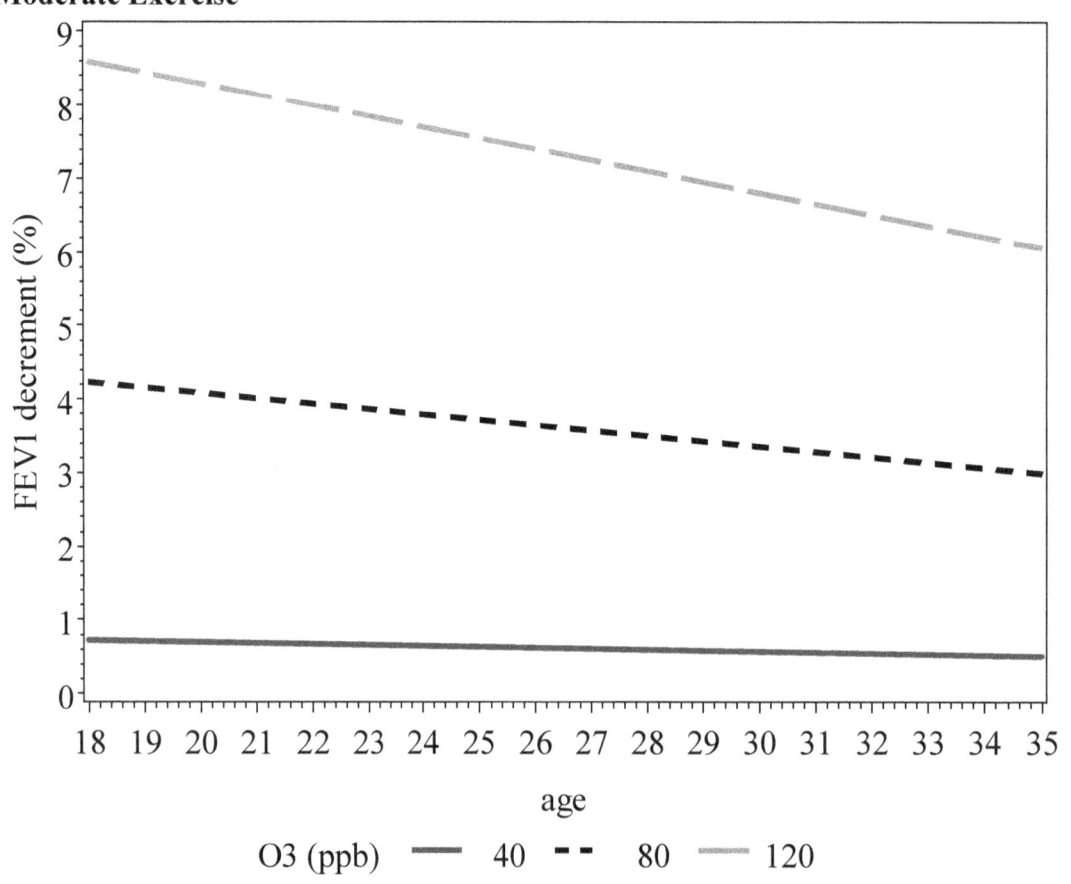

Figure 6E-5 shows the median response predicted by the McDonnell-Stewart-Smith model for a 6.6 hour exposure to 80 ppb ozone under intermittent moderate exercise (40 L/min, BSA=2 m^2). This was obtained by running the MSS model for different ages with the concentration and ventilation rate time series fixed, according to a typical protocol for a 6.6 hour exposure study. Subjects alternated 50 minutes of moderate exercise with 10 minutes of rest for the first three hours, with the exercise occurring first. For the next 35 minutes, subjects continued exposure at rest. For the remaining three hours of the exposure period, subjects again alternated 50 minutes of exercise with 10 minutes of rest.

Figure 6E-5. Median Response Predicted by the MSS Model For a 6.6 Hour Exposure to 80 ppb O$_3$ Under Intermittent Moderate Exercise (40 L/min, BSA=2 m^2)

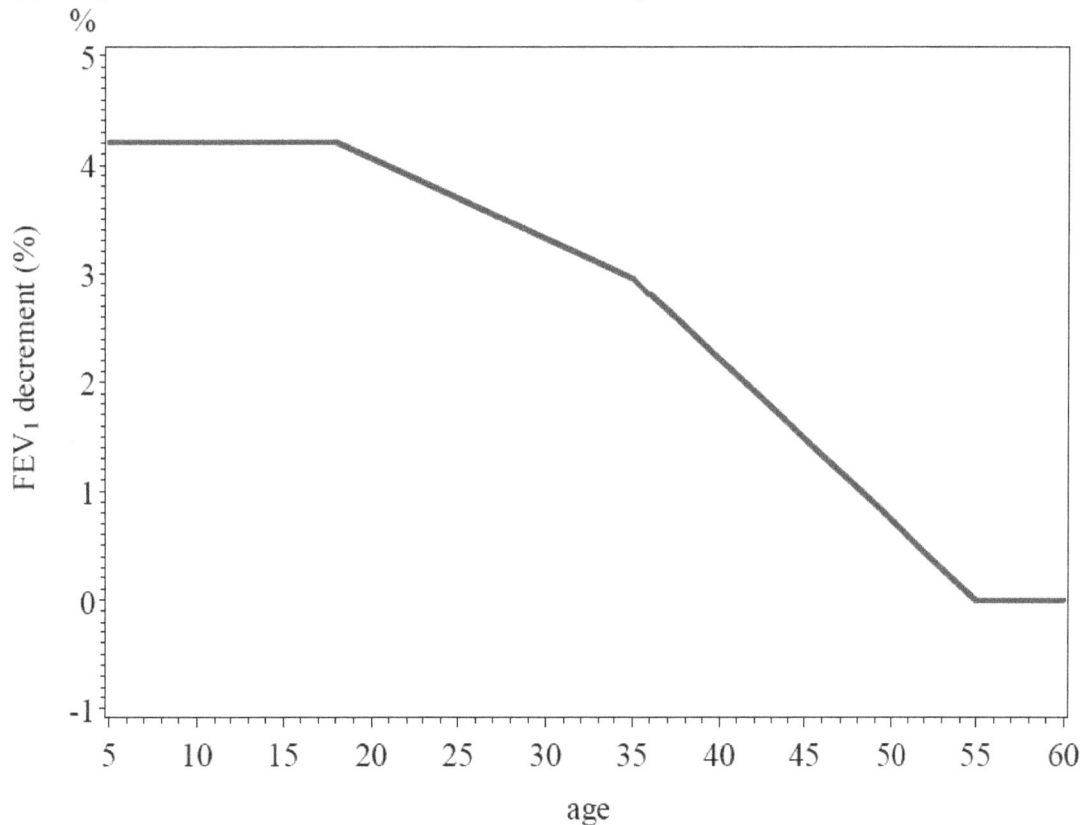

Physiology, Ventilation Rates, and Outdoor Exposures

The primary age-related factors in APEX of interest influencing estimated FEV_1 decrements (except the MSS model age term parameters) are given in Table 6E-2.

Table 6E-2. Selected factors in APEX which influence estimated FEV_1 decrements

Factor	Description
BSA	Body surface area is a basic physiological quantity which is an increasing function of age. Distributions of BSA that depend on age and sex are input to the APEX model.
RMR	Resting metabolic rate is a basic physiological quantity which depends on age. APEX models RMR using age- and sex-dependent distributions.
MET	Unadjusted MET distribution is a function only of the CHAD diary activity, which in turn depends on age. MET is a measure of the level of exertion during an activity.
MET_A	MET adjusted for excess post-exercise oxygen consumption (EPOC) is calculated in APEX as a function of unadjusted MET, RMR, MOXD, RecTime, VO_2max, BM, and ECF.
VE	Ventilation rate is an input to the MSS model. VE is calculated in APEX as a function of adjusted MET, VE regression coefficients, RMR, ECF, and BM.
EVR	Equivalent ventilation rate = VE/BSA is the term in the MSS model that incorporates all of the physiological variables.
Time outdoors	Time spent outdoors is a function only of the CHAD diary location, which in turn depends on age.
Exposure to ambient O_3 while outdoors	The ambient concentration levels while an individual is outdoors. This is related to age since it depends on the times of day and duration that an individual is outdoors, which depend on the CHAD diary activity patterns, which depend on age.

We find the hours of 2 pm to 9 pm are most relevant to exposures and activities leading to higher FEV_1 decrements, and have restricted this analysis to averages of quantities over these hours (data not shown). BSA, RMR, METmax, and VO_2max increase with age until around age 18 and then more or less level off. Looking at the 90th percentiles of the daily (hours of 2 pm to 9 pm) averages, we find that from ages 5 to 15, unadjusted MET decreases while adjusted MET increases. Ventilation rate (VE) increases with age from 5 to 15 while ventilation rate normalized by body surface area (EVR) decreases. Time spent outdoors is highest for ages 5 to 10 and around age 18, a result of the composition of the CHAD diaries. The ambient concentrations of O_3 exposed to while spending time outdoors is highest for ages 5 to 15. The four most influential of these factors on the relationship of FEV_1 decrement with age are the decreasing adjusted MET and decreasing EVR (with increasing age), the high time spent outdoors, and the high exposure concentration while outdoors. These all lead to children having higher FEV_1 decrements than adults.

The graphs of boxplots by age on the following pages show the distribution of the factor for each age. The boxes indicate the 25th and 75th percentiles, the midlines are the medians, and the whiskers extend to the minima and maxima. The graphs of single lines are the 90th percentiles for each age. The factors are daily averages over hours 14 to 21 from an APEX simulation of the Atlanta metropolitan area for April-June 2006 (10,000 profiles simulated, ages 5 to 35).

VE is an increasing function of the exertion level as measured by MET_A (among other factors). We see that while younger children (within the 5-18 age range) have higher CHAD activity levels as measured by MET, they have lower exertion levels as measured by MET_A. This is primarily due to smaller maximum MET levels, smaller maximum VO_2, and smaller MOXD for the younger children, which result in lower ventilation rates. Normalization of VE by BSA results in higher levels of EVR for the younger children. It is this factor (EVR) which drives the trend for children we see in Figure 6E-2.

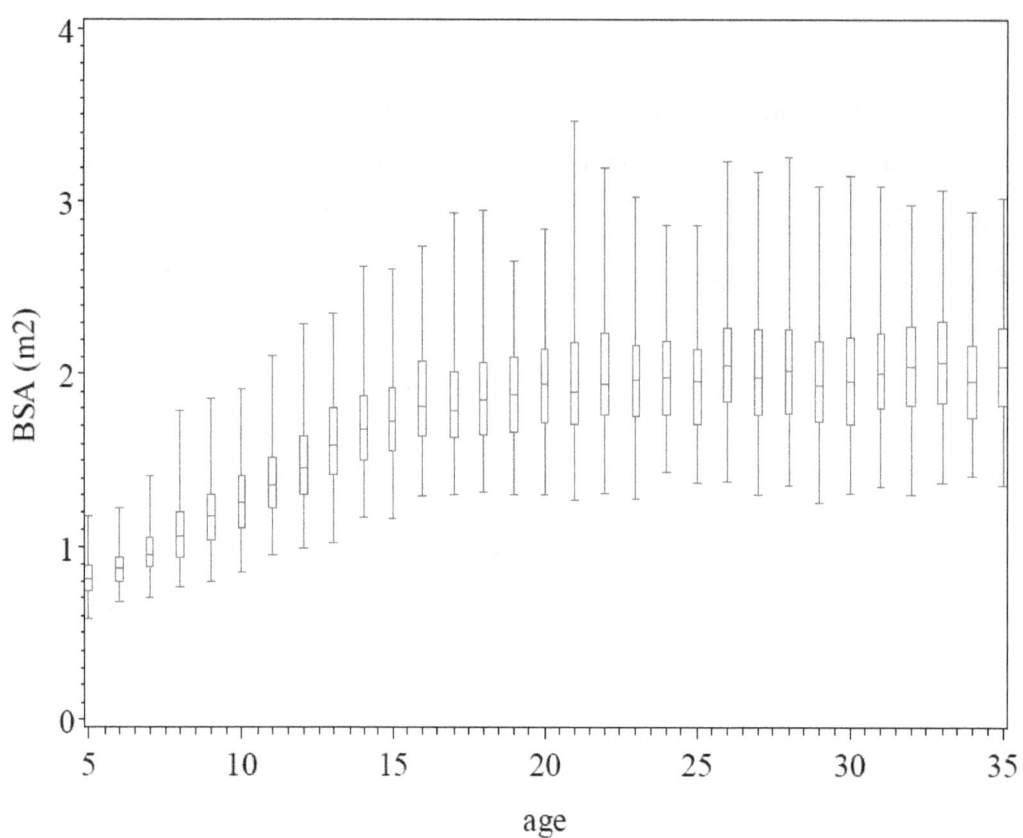

Figure 6E-6. Distribution of Daily Average (hours 14-21) BSA (m²) vs. age

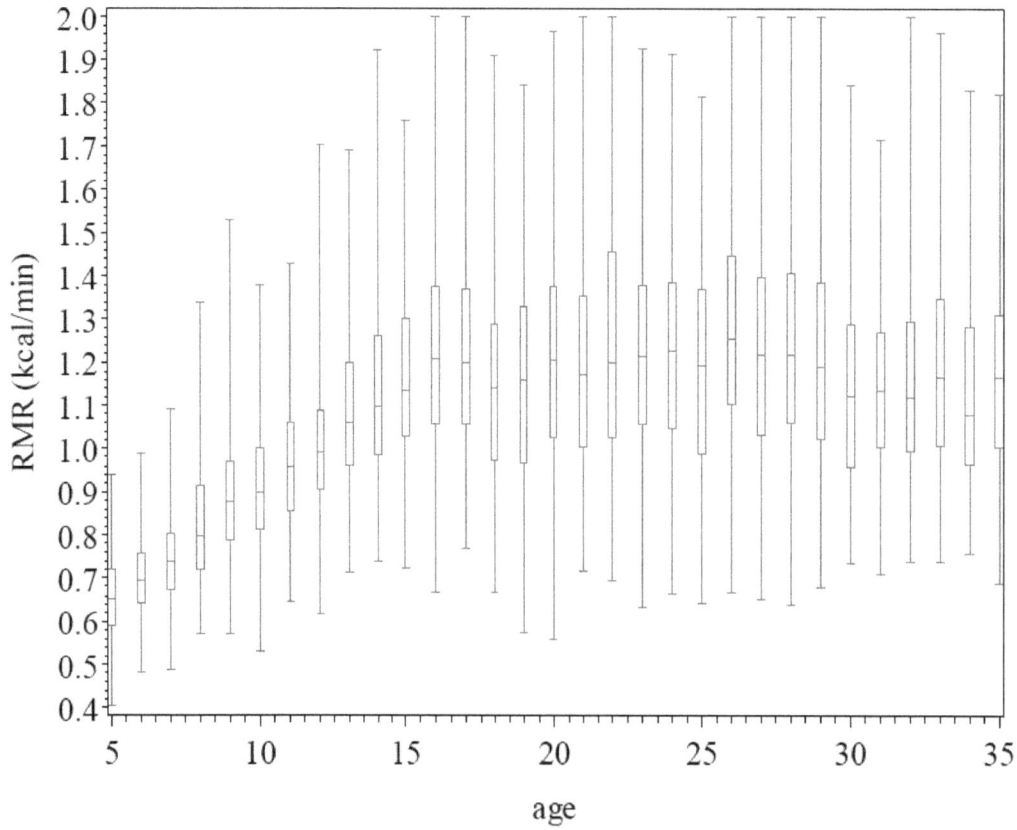

Figure 6E-7. Distribution of Daily Average (hours 14-21) RMR (kcal/min) vs. age

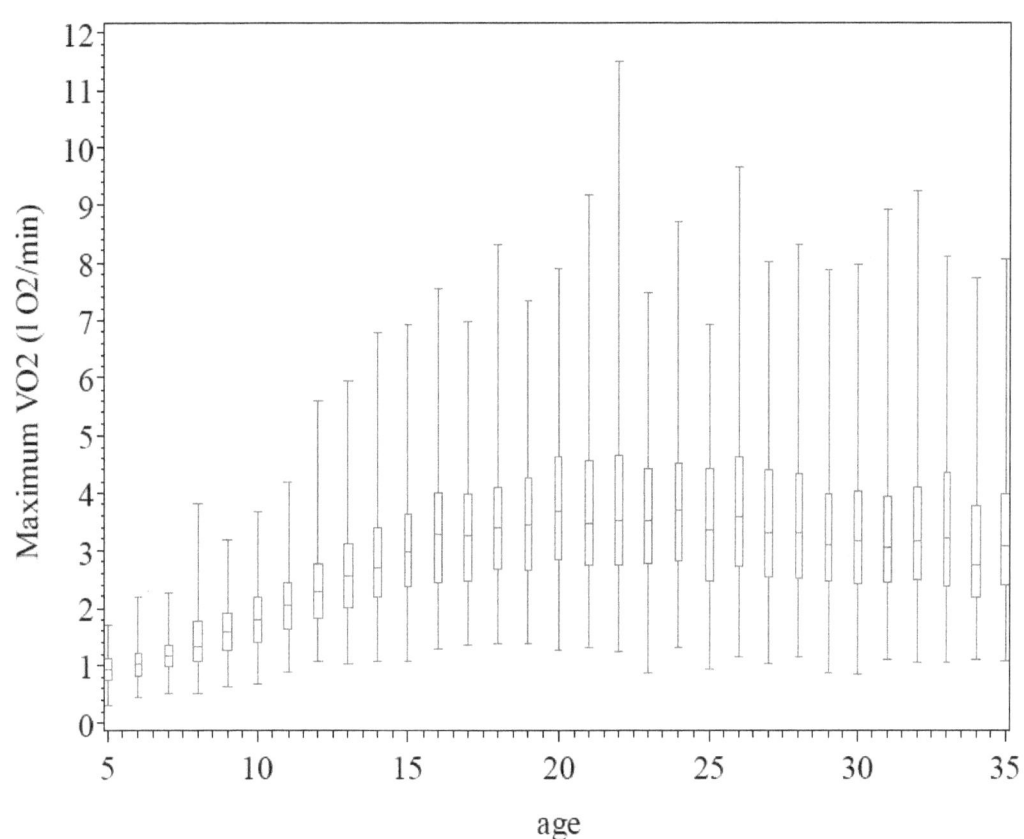

Figure 6E-8. Distribution of Daily Average (hours 14-21) Maximum VO₂ (L O₂/min) vs. age

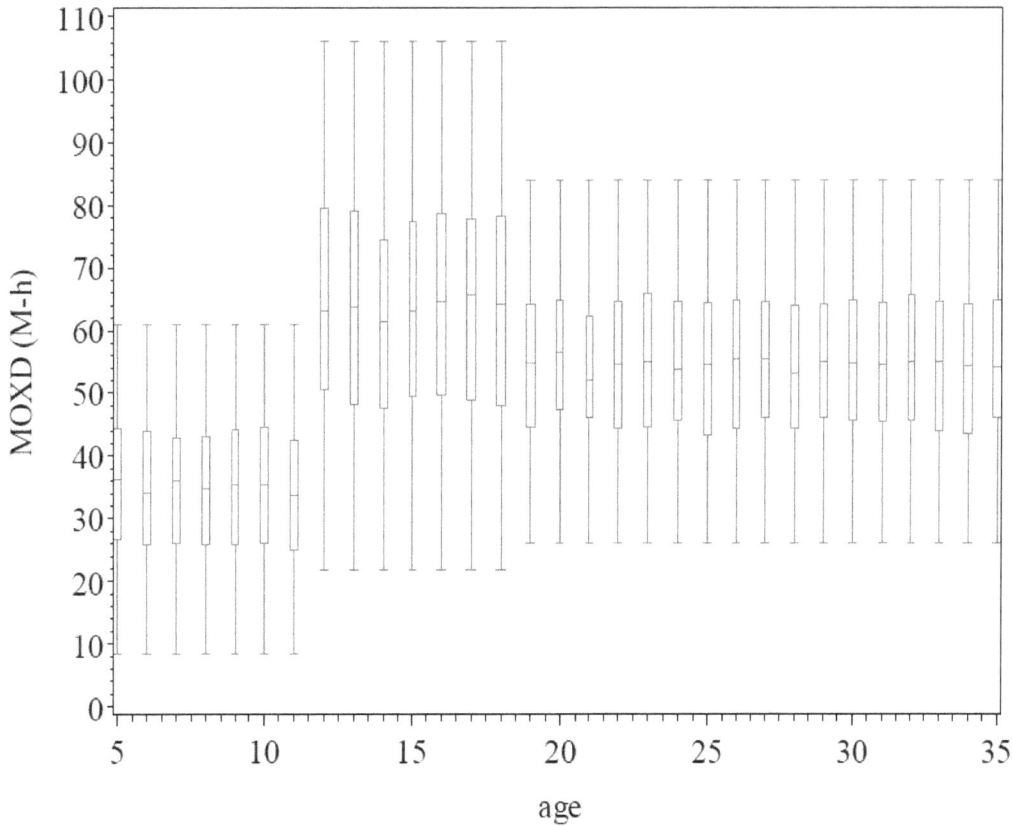

Figure 6E-9. Distribution of Daily Average (hours 14-21) MOXD (M-h) vs. age

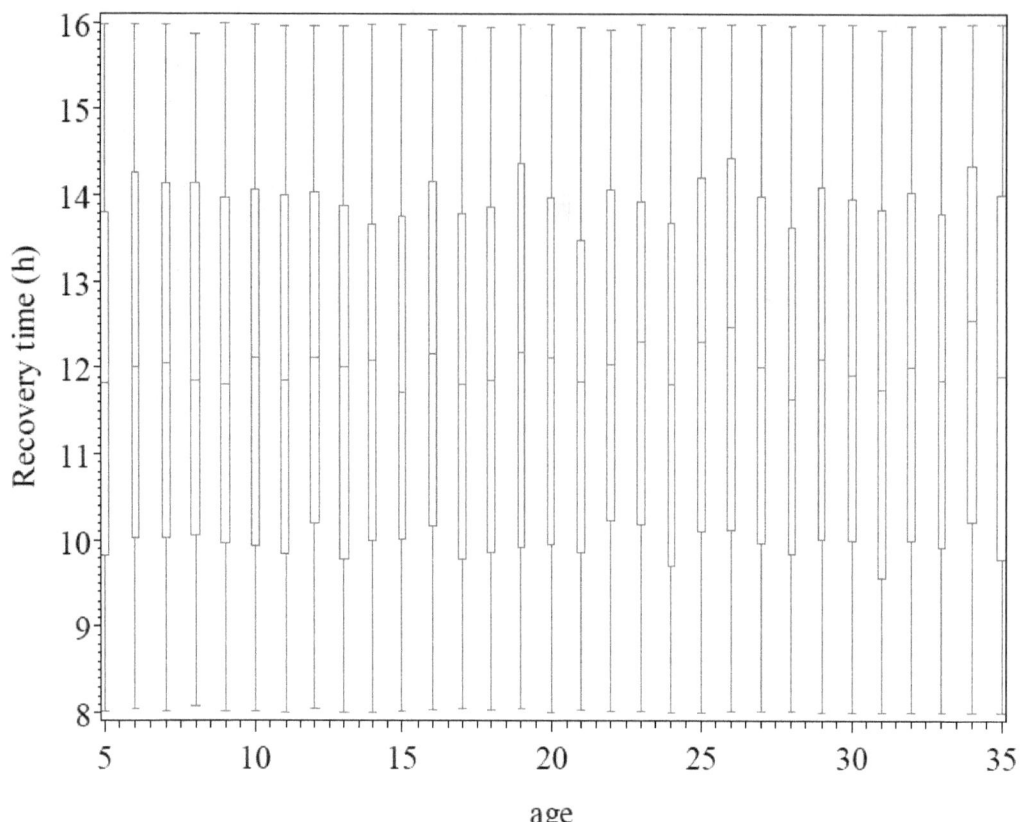

Figure 6E-10. Distribution of Daily Average (hours 14-21) Recovery Time (hours) vs. age

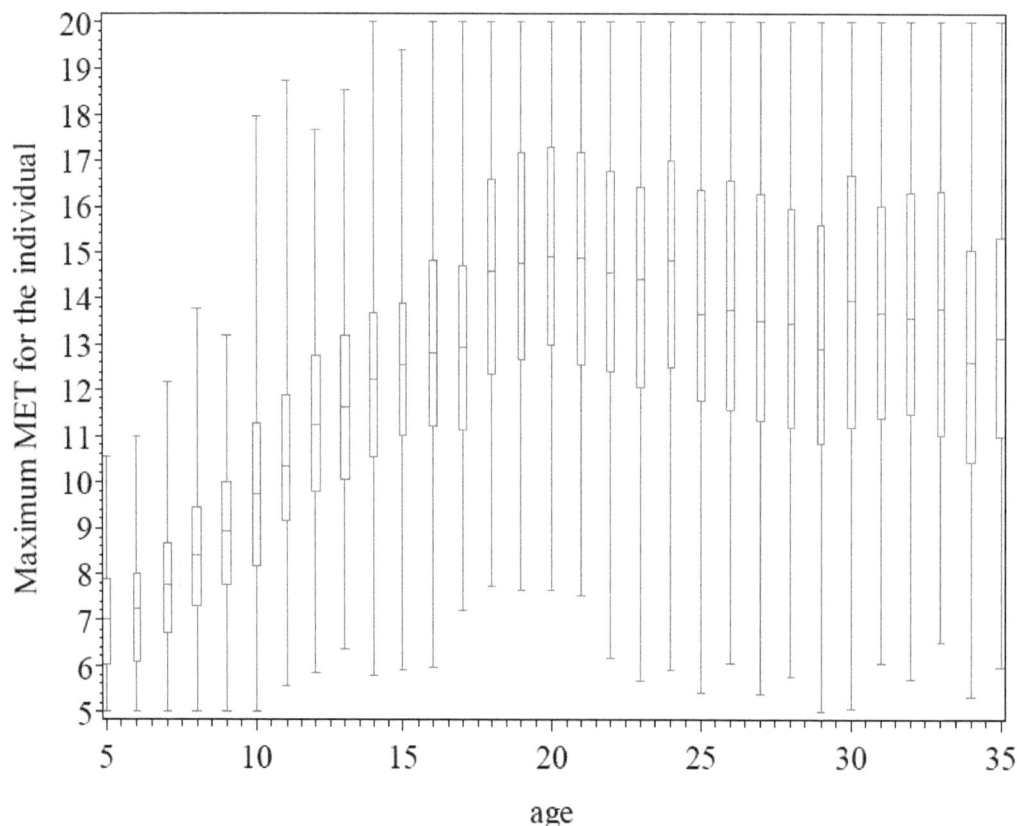

Figure 6E-11. Distribution of Maximum MET vs. age

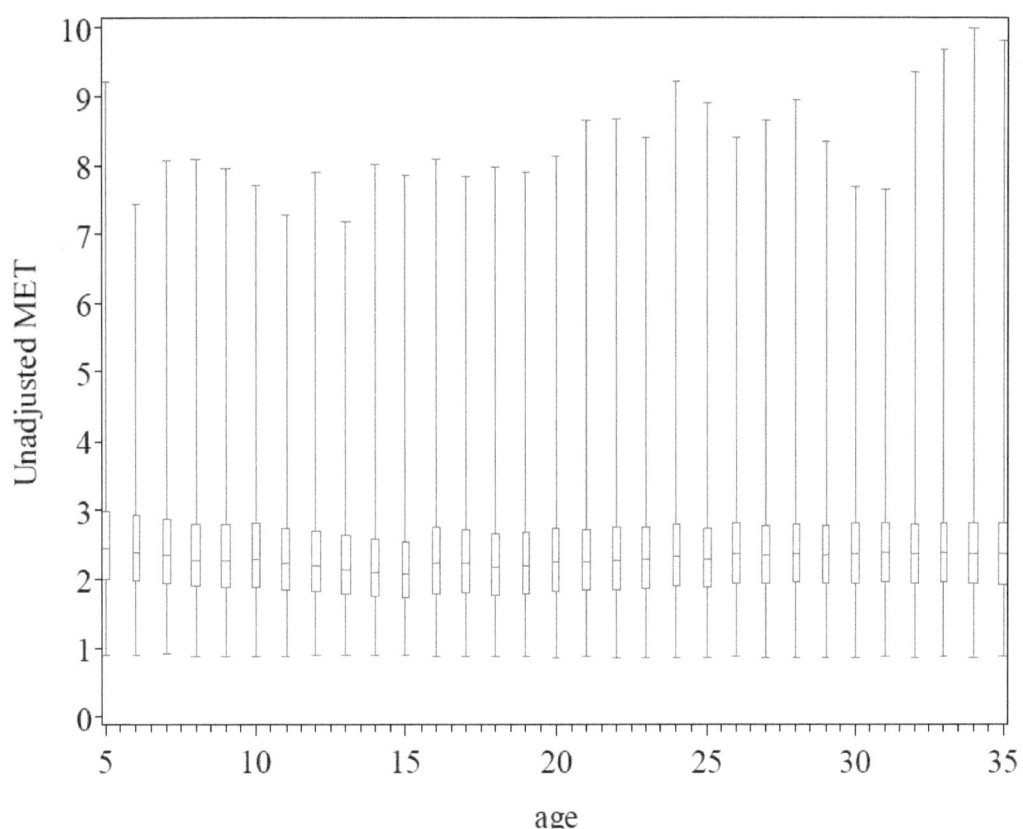

Figure 6E-12. Distribution of Daily Average (hours 14-21) Unadjusted MET vs. age

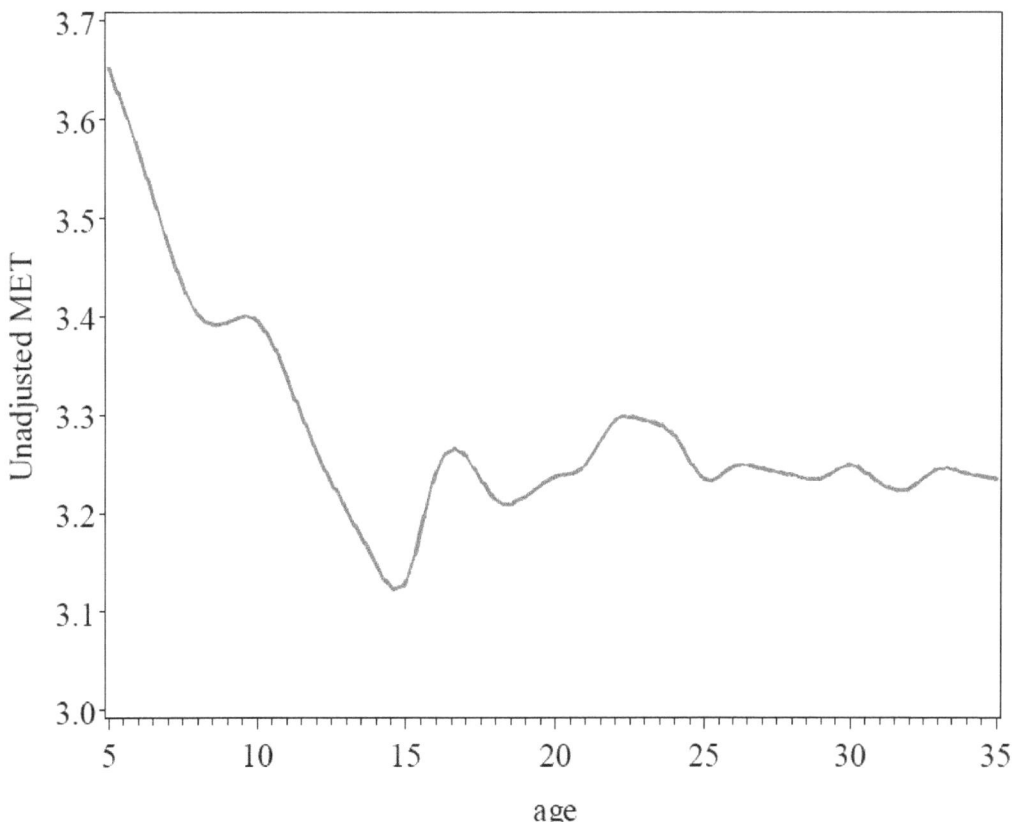

Figure 6E-13. Daily Average (hours 14-21) Unadjusted MET (90th percentiles) vs. Age

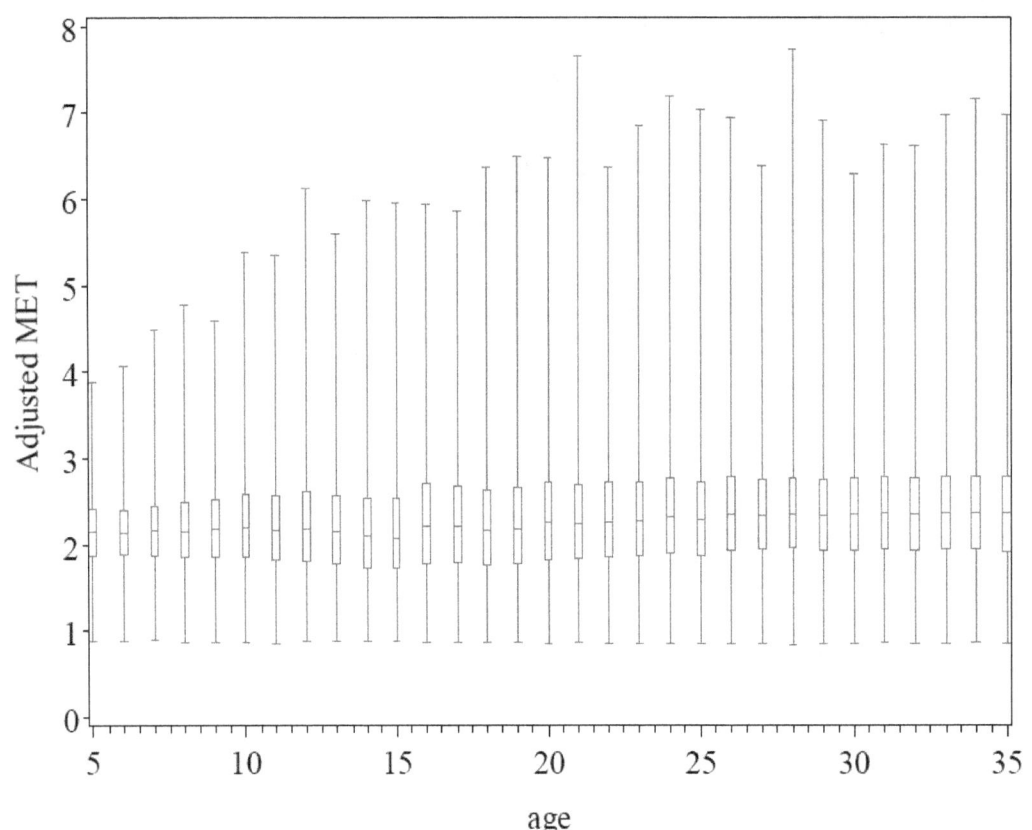

Figure 6E-14. Distribution of Daily Average (hours 14-21) Adjusted MET vs. age

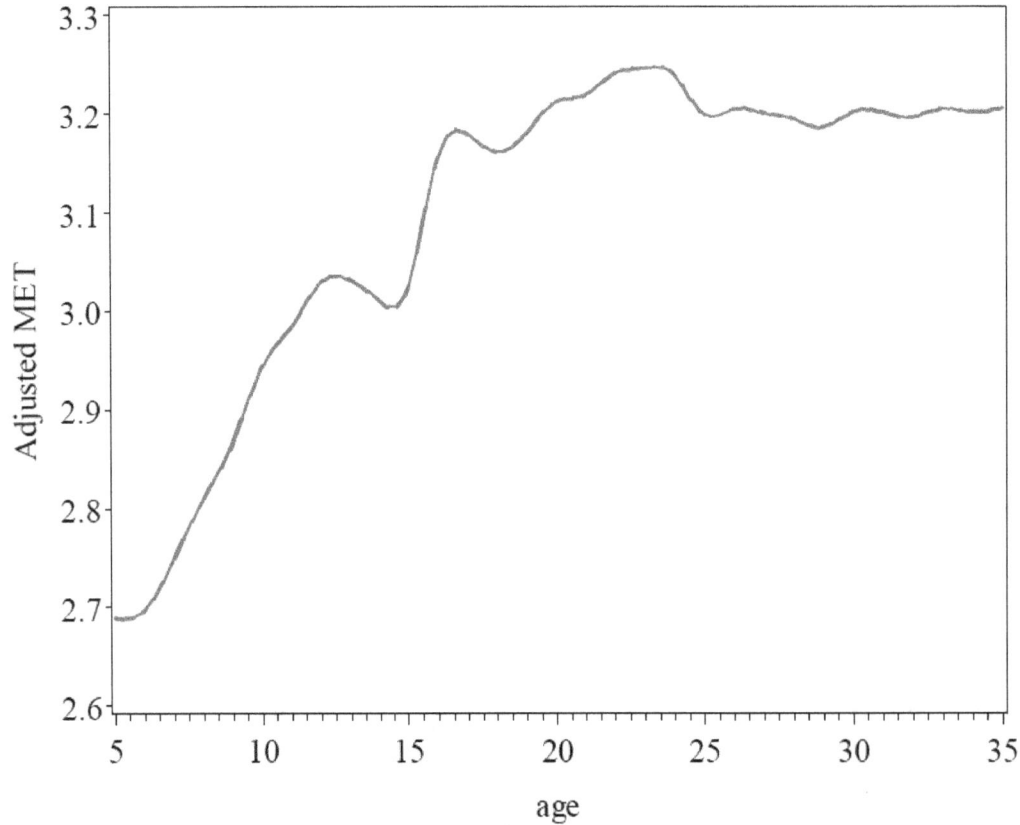

Figure 6E-15. Daily Average (hours 14-21) Adjusted MET (90th percentiles) vs. Age

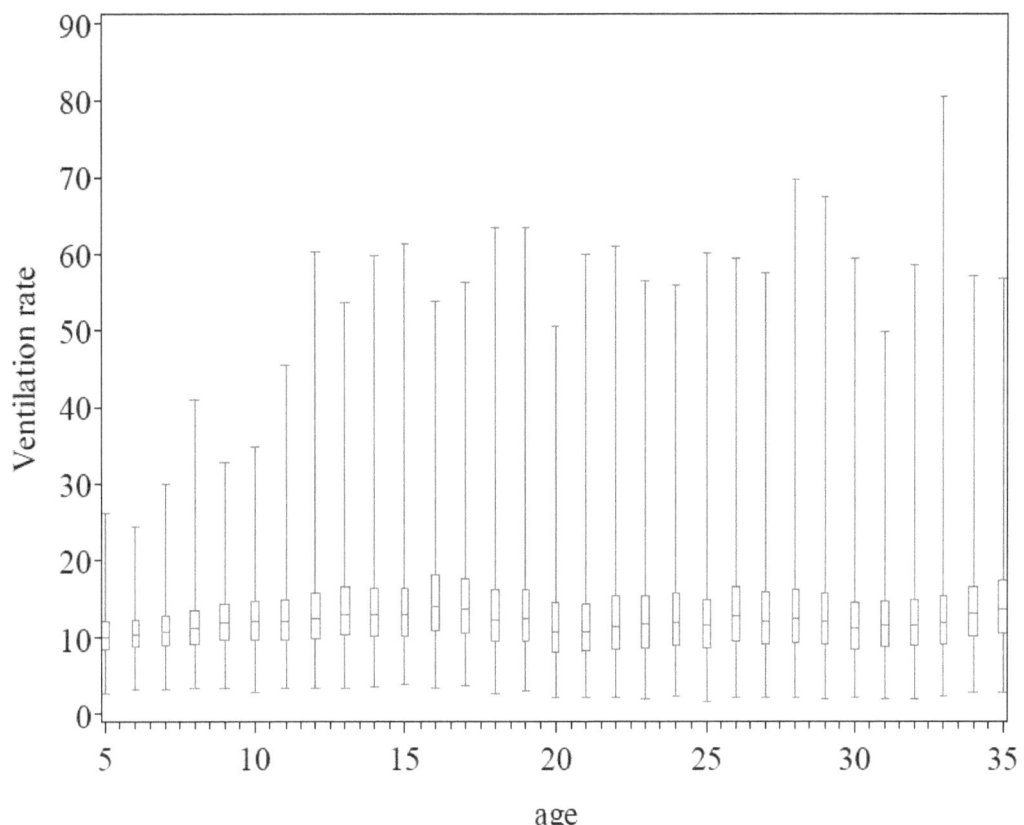

Figure 6E-16. Distribution of Daily Average (hours 14-21) Ventilation Rate (L/min) vs. age

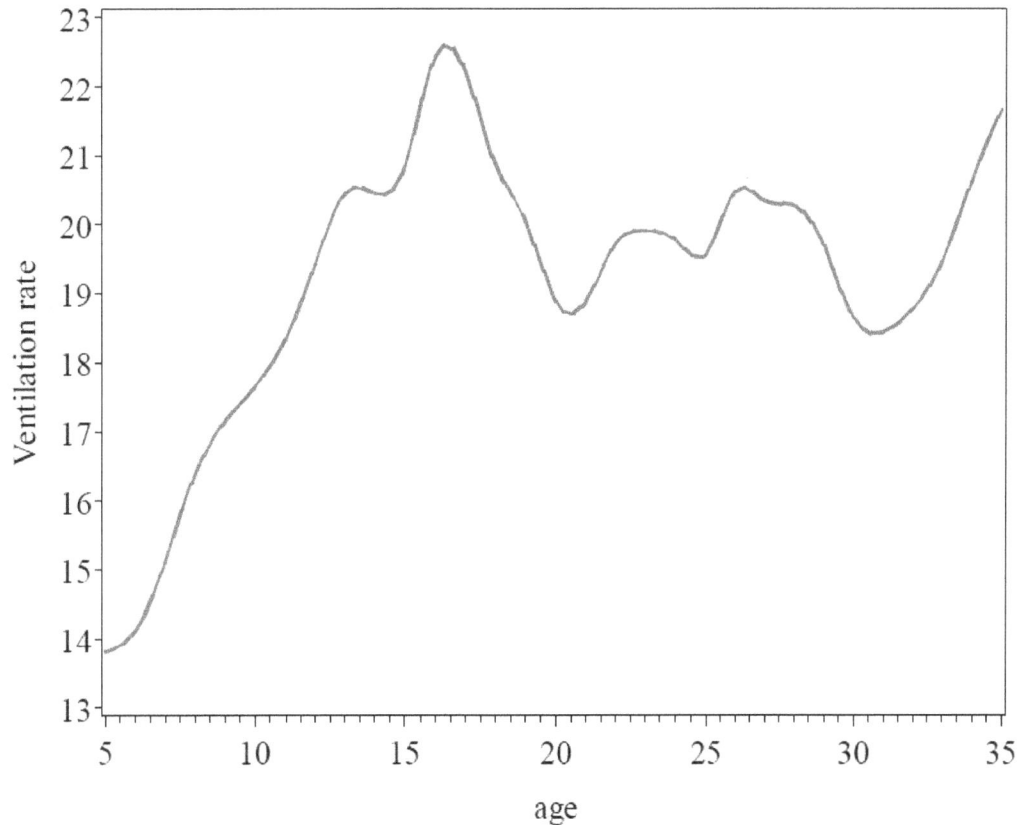

Figure 6E-17. Daily Average (hours 14-21) Ventilation Rate (L/min) (90th percentiles) vs. Age

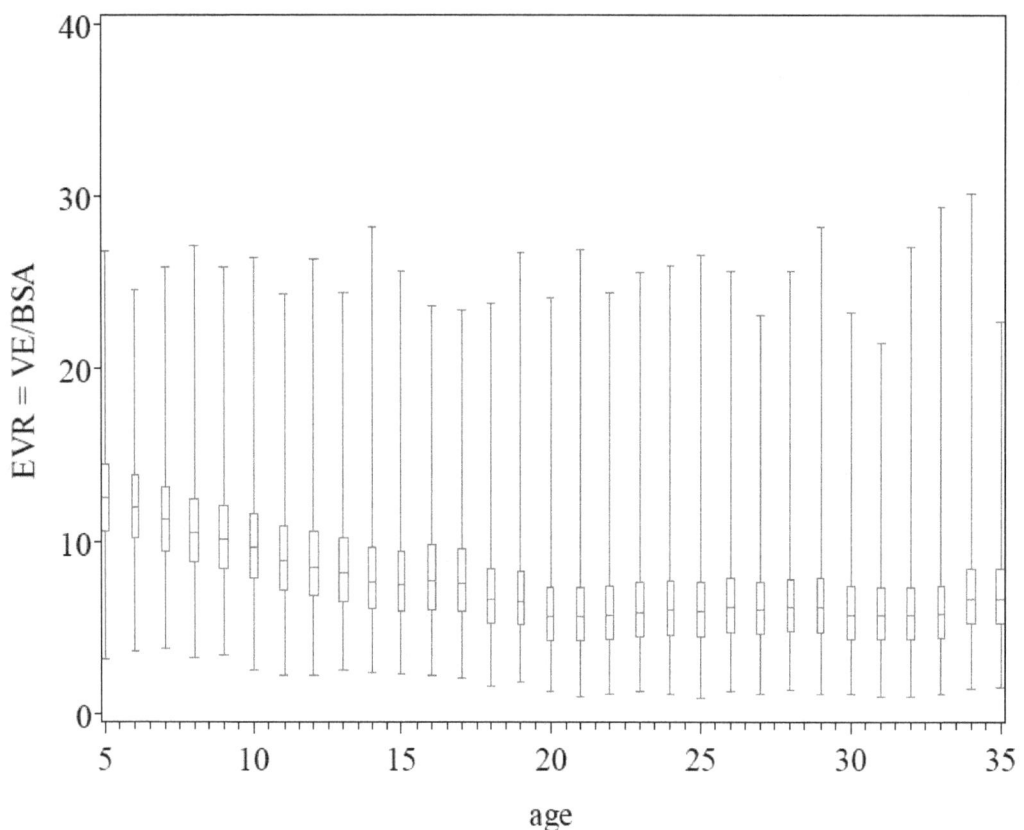

Figure 6E-18. Distribution of Daily Average (hours 14-21) EVR (L/min/m²) vs. age

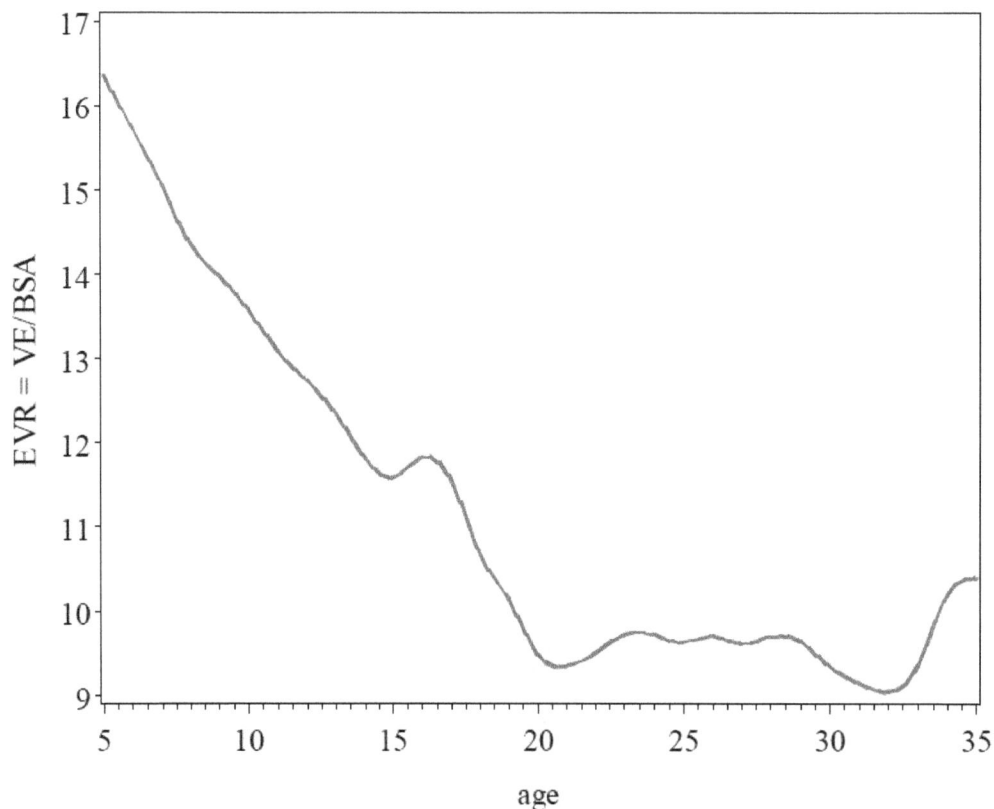

Figure 6E-19. Daily Average (hours 14-21) EVR (L/min/m²) (90th percentiles) vs. Age

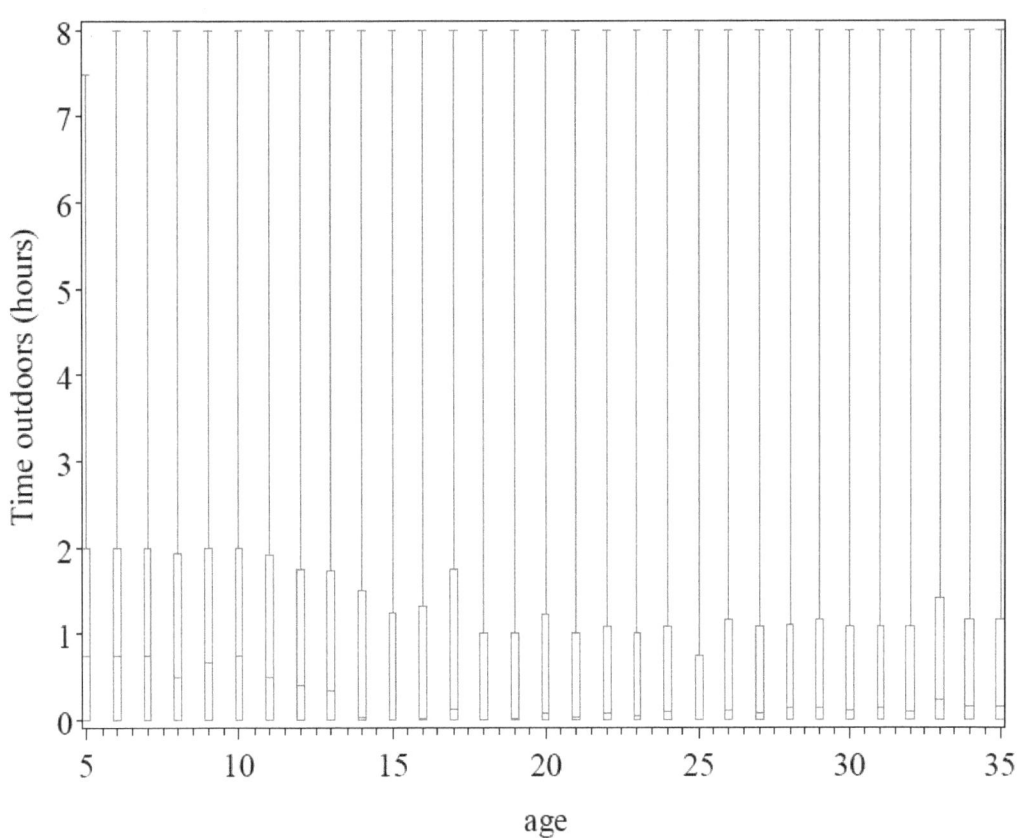

Figure 6E-20. Distribution of Daily Average (hours 14-21) Time Outdoors (hours) vs. age

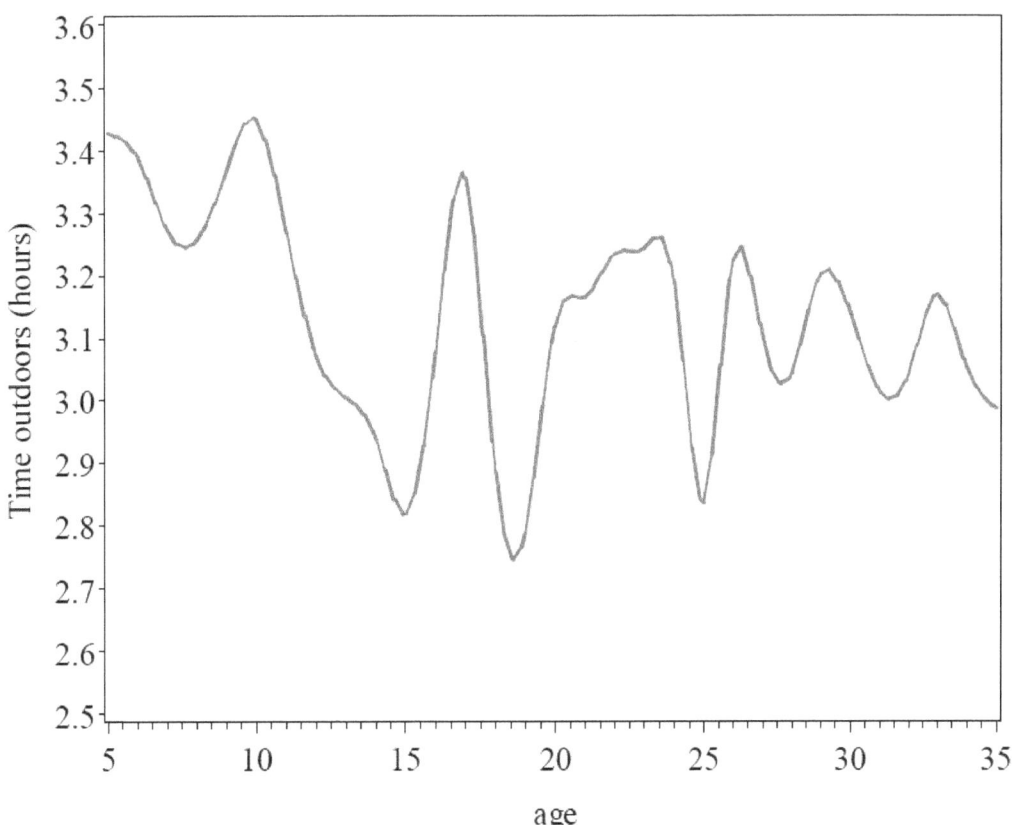

Figure 6E-21. Daily Average (hours 14-21) Time Outdoors (hours) (90th percentiles) vs. Age

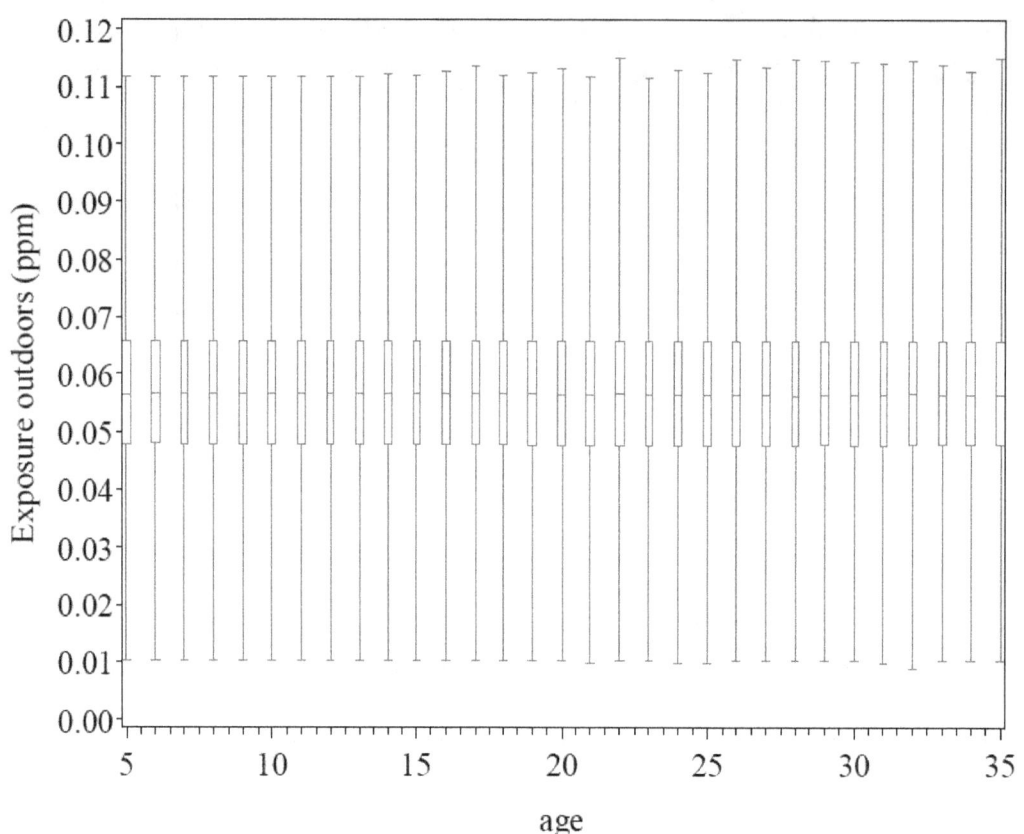

Figure 6E-22. Distribution of Daily Average (hours 14-21) Exposure Outdoors (ppm) vs. age

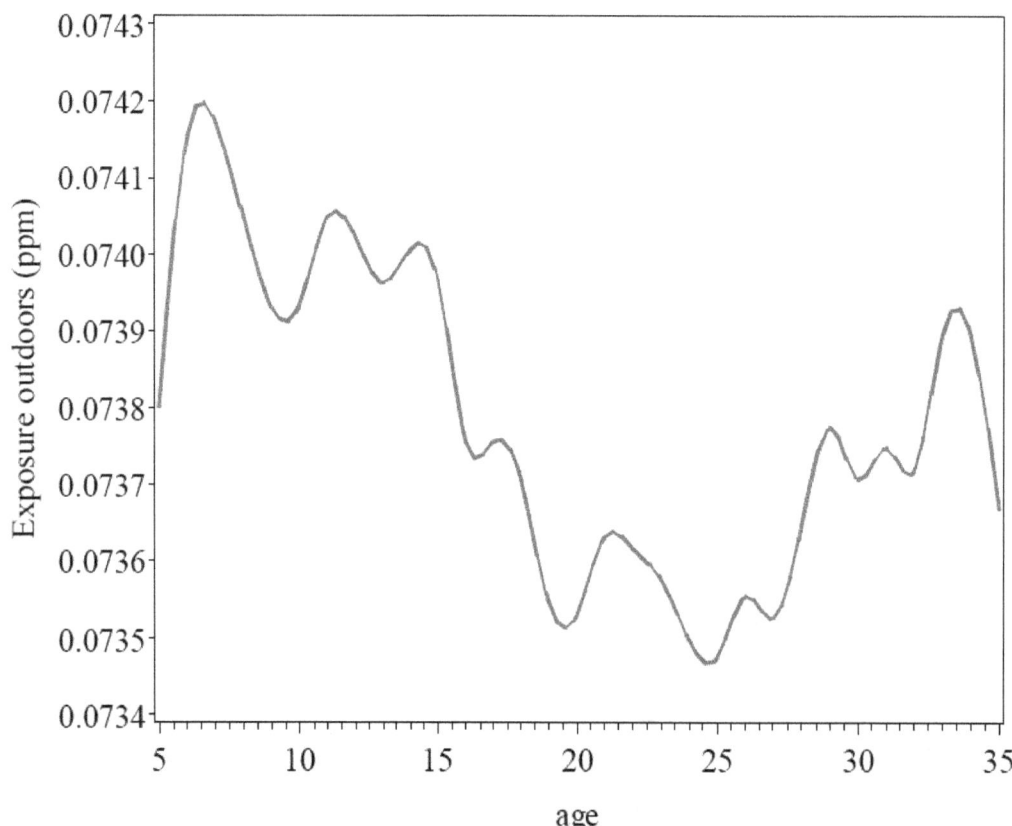

Figure 6E-23. Daily Average (hours 14-21) Exposure Outdoors (ppm) (90th percentiles) vs. Age

6E-18

1   **Alternative Age Term For Children**

2   The results of an APEX simulation of the Atlanta 2006 base case (March 1 – October 30)

3   using an alternative age term for children are presented in Table 6E-3. This age term is based on

4   the assumption that the responsiveness of children to ozone is about the same as for young adults

5   (ISA, 2012, p. 6-21) and the age term is set to the average age term over ages 18 to 35 (α_2 = 2.7,

6   see Table 6-2). The alternative age term results in estimates that are lower, but not dramatically

7   so. Most of the age effects are due to the factors discussed in the previous section.

8   **Table 6E-3. Percents of the population ages 5 to 18 with one or more days during the ozone**
9   **season with lung function (FEV$_1$) decrements more than 10, 15, and 20% (Atlanta 2006**
10   **base case). MSS threshold model and MSS threshold model with alternative age term.**

Model	$\Delta FEV_1 \geq$ 10%	$\Delta FEV_1 \geq$ 15%	$\Delta FEV_1 \geq$ 20%
MSS	31%	13%	6.4%
MSS, alternative age term	27%	10%	4.6%

11

12

13

1
2

1

2

3

Appendix 6F

4

5

MSS Model Variability Term Sensitivities

6

7

8

1   **Comparison of APEX Simulation Results Using Monitors and Tracts Air Quality**

2   In the first draft REA, monitor-level air quality was provided as input to the APEX

3   model. As discussed in Chapter 5, tract-level air quality was used in APEX for this second draft

4   REA. Monitor-level air quality is used for the APEX simulations in this section, since these

5   simulations take less time to run. This does not affect the analyses here, since the two air quality

6   formats yield very similar results, as can be seen by comparing Table 6F-1 with Table 6F-2.

7   Tables 6F-1 through 6F-3 are based on Atlanta 2006 base case APEX simulations of

8   200,000 individuals, 52,436 of which are children ages 5 to 18. These simulations all used the

9   same random number seed to hold all variables constant except for the sensitivity variables.

10   Table 6F-4 is based on Atlanta 2006 base case APEX simulations of 50,000 individuals, all

11   children ages 5 to 18.

12   **Table 6F-1. Percents of the population by age group with one or more days during the**
13   **ozone season with lung function (FEV₁) decrements more than 10, 15, and 20% (Atlanta**
14   **2006 base case). MSS Threshold model, monitors air quality.**

Age Group	$\Delta FEV_1 \geq$ 10%	$\Delta FEV_1 \geq$ 15%	$\Delta FEV_1 \geq$ 20%
5 to 18	31%	13%	6.4%
19 to 35	11%	3.1%	1.3%
36 to 55	3.7%	0.60%	0.14%

15

16   **Table 6F-2. Percents of the population by age group with one or more days during the**
17   **ozone season with lung function (FEV₁) decrements more than 10, 15, and 20% (Atlanta**
18   **2006 base case). MSS Threshold model, tracts air quality.**

Age Group	$\Delta FEV_1 \geq$ 10%	$\Delta FEV_1 \geq$ 15%	$\Delta FEV_1 \geq$ 20%
5 to 18	30%	12%	5.7%
19 to 35	11%	2.9%	1.1%
36 to 55	3.3%	0.50%	0.12%

19

1   **MSS Model Variability Term Sensitivities**

2   The variability term has a Gaussian distribution with mean zero and variance 17.1 (in the

3   threshold model) and is sampled daily for each simulated individual. Since the actual values are

4   bounded, we have truncated the variability term distribution at ±2 standard deviations (±8.27), a

5   convention we use for the distributions of several physiological variables input to APEX in the

6   physiology input file. To look at the effect of truncating the variability distribution, we

7   conducted simulations using the threshold model with this constraint removed. The results of

8   this simulation are given in Table 6F-3.

9   We see that this constraint has a very large effect on the results for percents of the

10   population with FEV_1 decrements $\geq$ 10 and 15% and less of an effect for 20%. When fitting the

11   MSS model, the actual values of the variability term range from about -20 to 20. If we truncate

12   the variability term distribution at ±12, 16, and 20, then the corresponding simulation results for

13   ages 18 to 35 are given in Table 6F-4.

14   Clearly the variability term in the MSS model is a key parameter and is influential in

15   predicting the proportions of the population with FEV_1 decrements > 10, and 15%. The

16   assumption that the distribution of this term is Gaussian is convenient for fitting the model, but is

17   not accurate. The extent to which this mis-specification affects the estimates of the parameters

18   of the MSS model and its predictions is not clear.

19

20   **Table 6F-3. Percents of the population by age group with one or more days during the**
21   **ozone season with lung function (FEV_1) decrements more than 10, 15, and 20% (Atlanta**
22   **2006 base case). MSS Threshold model, monitors air quality, non-truncated Gaussian**
23   **variability term.**

Age Group	$\Delta FEV_1 \geq$ 10%	$\Delta FEV_1 \geq$ 15%	$\Delta FEV_1 \geq$ 20%
5 to 18	92%	19%	7.1%
19 to 35	88%	7.5%	1.4%
36 to 55	87%	4.6%	0.22%

24

25

1 **Table 6F-4. Percents of the population ages 5 to 18 with one or more days during the ozone**
2 **season with lung function (FEV$_1$) decrements more than 10, 15, and 20% (Atlanta 2006**
3 **base case). MSS Threshold model, monitors air quality, variability term truncated at ±12,**
4 **16, 20, and no truncation limit.**

ε_{ijk} limit	$\Delta FEV_1 \geq$ 10%	$\Delta FEV_1 \geq$ 15%	$\Delta FEV_1 \geq$ 20%
±8.27	38%	14%	6.8%
±12	92%	15%	7.1%
±16	92%	19%	7.1%
±20	92%	19%	7.1%
no limit	92%	19%	7.1%

5

6

United States
Environmental Protection
Agency

Office of Air Quality Planning and Standards
Air Quality Strategies and Standards Division
Research Triangle Park, NC

Publication No. EPA-452/P-14-004d
February 2014